Green Energy and Technology

Climate change, environmental impact and the limited natural resources urge scientific research and novel technical solutions. The monograph series Green Energy and Technology serves as a publishing platform for scientific and technological approaches to “green”—i.e. environmentally friendly and sustainable—technologies. While a focus lies on energy and power supply, it also covers “green” solutions in industrial engineering and engineering design. Green Energy and Technology addresses researchers, advanced students, technical consultants as well as decision makers in industries and politics. Hence, the level of presentation spans from instructional to highly technical.

**Indexed in Scopus**.

More information about this series at http://www.springer.com/series/8059

Nnamdi Nwulu · Saheed Lekan Gbadamosi

# Optimal Operation and Control of Power Systems Using an Algebraic Modelling Language

Springer

Nnamdi Nwulu
Department of Electrical
and Electronic Engineering Science
University of Johannesburg
Johannesburg, South Africa

Saheed Lekan Gbadamosi
Department of Electrical
and Electronic Engineering Science
University of Johannesburg
Johannesburg, South Africa

ISSN 1865-3529 ISSN 1865-3537 (electronic)
Green Energy and Technology
ISBN 978-3-030-00394-4 ISBN 978-3-030-00395-1 (eBook)
https://doi.org/10.1007/978-3-030-00395-1



This Springer imprint is published by the registered company Springer Nature Switzerland AG
The registered company address is: Gewerbestrasse 11, 6330 Cham, Switzerland

# Preface

Today's modern grids are complex interconnected systems with increased customer participation and intricate ancillary services. The incorporation of modern technologies including renewable energy sources and electric vehicles have increased the complexity of modern power systems. Furthermore, today's power system requires high reliability as brownouts and blackouts have tremendous economic, social, environmental, and political implications. This has led to increased focus on how to optimally operate, monitor and control power systems. Thus, this book *Optimal Operation and Control of Power Systems Using an Algebraic Modelling Language* focuses on the use of Algebraic Modelling Languages as a simplified approach for modelling and solving complex mathematical optimisation problems found in modern power grids. The Advanced Interactive Multidimensional Modelling System (AIMMS) is the AML used in this book.

This comprehensive text for power system operations is divided into three parts. The first part is an introductory part and spans Chaps. 1–3.

In Chap. 1, an introduction and overview of the electrical power system is given. It covers the major components of a power system and reviews the various renewable energy sources (RES) that constitute today's energy mix. The chapter also delves into smart grids and considers various definitions of a smart grid. The deregulation of the electric power industry across nations of the world is also considered and the chapter concludes with operational and control issues in a power system.

Chapter 2 reviews the concept of demand side management (DSM) programs, which aim to foster a more responsible use of electrical energy. DSM encompasses both energy efficiency and demand response (DR) programs. Examples of energy efficiency programs (a retrofit program), price-based DR (time of use appliance scheduling) and IB-DR programs (game theory-based incentive-based DR) were discussed in detail and their mathematical formulations given.

In Chap. 3, mathematical optimization modelling is considered. The two broad classifications of mathematical modelling solution approaches, classical and meta-heuristic methods, are detailed and various examples are given. The chapter also gives practical steps on how to model mathematical problems. Furthermore, algebraic modelling languages (AMLs) are considered and AIMMS, which is a prominent AML, is introduced. Finally, examples were shows of different energy efficiency and DR programs and their notation in AIMMS.

The second part details various power system operations and control models including demand response programs and spans Chaps. 4–10. The AIMMS source code for all the models is included.

Chapter 4 considers the Dynamic Economic Emissions Dispatch (DEED) problem which is concerned with assigning customer loads to the available power-generating units in order to ensure a reliable and secured system whilst minimizing fuel cost and emissions. In this chapter, different variations of the DEED problems were investigated. The variations considered include the Dynamic Economic Dispatch (DED), Emissions Constrained Dynamic Economic Dispatch (ECDED), Bid Based Dynamic Economic Emissions Dispatch (BBDEED) and the Profit Based Dynamic Economic Emissions Dispatch (PBDEED). An Incentive Based DR program is also incorporated into the DEED problem-giving rise to the IBDEED problem and is considered in this chapter. All models were solved using AIMMS with source codes given.

In Chap. 5, the Generator Maintenance Scheduling problem was presented. Specifically, the Reliability Constrained GMS (RC-GMS), which seeks to minimize the sum squares of reserve of the generating units without violating maintenance window constraints is developed and solved using AIMMS.

Chapter 6 considers the Combined Heat and Power Dispatch (CHPD), which obtains the optimal dispatch of combined heat and power generators whilst minimizing daily operational cost and satisfying users power and heat demand throughout the total dispatch interval. Variants like the Combined Heat and Power Dynamic Economic Dispatch (CHPDED), Combined Heat and Power Pure Dynamic Emissions Dispatch (CHPPDED), Combined Heat and Power Dynamic Economic Emissions Dispatch (CHPDEED) and Profit-based Combined Heat and Power Dynamic Economic Emissions Dispatch (PBCHPDEED) are considered. Again, an incentive-based DR program is integrated into the CHPDEED problem giving rise to the IB-CHPDEED formulation. All models are solved using AIMMS and the source code given.

In Chap. 7, a grid-connected hybrid renewable energy powered microgrid incorporating a DR program is presented. The main aim is to minimize the operational cost of the fossil fuel generators and the cost of transferrable power traded between the main grid and microgrid, as well as maximizing the incentive based DR. Another variant of the model but including battery storage systems is also detailed. Both models are modelled and solved with AIMMS with relevant source codes detailed.

Chapter 8 considers the Optimal Power Flow whose objective is to minimize the operational cost of the power system while maintaining effective generation scheduling with efficient power flow on the transmission lines. The Direct Current OPF is the variant of OPF modelled in this chapter and the AIMMS code and results detailed.

In Chap. 9, the Transmission Network Expansion Planning (TNEP) is presented. TNEP aims at determining the optimal approach for adequate selection of transmission lines to be constructed to meet the growing energy demand, as well as facilitating power transfer from far-away locations to consumers. Two variants of TNEP: The static and dynamic TNEP are modelled and the incorporation of demand response programs into the TNEP problem is presented with the relevant AIMMS code detailed.

Chapter 10 gives detail of the GCEP problem formulation. The chapter considers different planning approaches, ranging from load forecasting, real and reactive power capacity to DR integrated GCEP. The GCEP models were formulated for both DC and AC planning methods, as well as real and reactive power capacity. Different examples are illustrated in this chapter and their solutions are provided with appropriate AIMMS codes.

The final part consists only of Chap. 11, which details the conclusions of the book whilst highlighting possible future extensions and considerations.

Johannesburg, South Africa

Nnamdi Nwulu
Saheed Lekan Gbadamosi

# Contents

# List of Figures

# List of Tables

# Chapter 1
# Power Systems and Renewable Energy Systems

## 1.1 Introduction and Overview

Electric power systems, sometimes simply referred to as power systems, can be defined as a network of electrical devices that are used to generate, transmit, and distribute electrical power.

From this definition, it can be stated that electric power systems have three key roles: the generation, transmission, and distribution of electrical power. Thus, power systems can be classified using these three critical categories. Figure 1.1 gives the basic components of a power system.

### *1.1.1 Generation of Electric Power*

The generation of electrical power or electricity occurs via the conversion of mechanical energy to electrical energy. The devices used for this are often referred to as electromechanical generators. They work by converting kinetic energy and forcing a generator to rotate, thus generating electric power.

They can be powered either by thermal processes or by the flow of water or wind. Thermal processes typically use coal and natural gas as the primary fuel. Electricity is typically generated at low voltage and stepped up to be transmitted using the transmission lines.

Most of the thermal-powered energy sources described above are referred to as non-RES. They are termed non-renewable because they cannot be replenished in a manner commensurate with consumption. Non-RES are typically fossil fuels. Fossil fuels are those formed from the remains of organisms that lived thousands of years ago. These fuels are a large contributor to harmful atmospheric emissions. RES are energy sources that can be replenished in a manner commensurate with consumption. These energy sources do not release harmful atmospheric emissions and thus are increasingly being used to generate electrical energy. Examples of this kind



N. Nwulu and S. L. Gbadamosi, *Optimal Operation and Control of Power Systems Using an Algebraic Modelling Language*, Green Energy and Technology,
https://doi.org/10.1007/978-3-030-00395-1_1

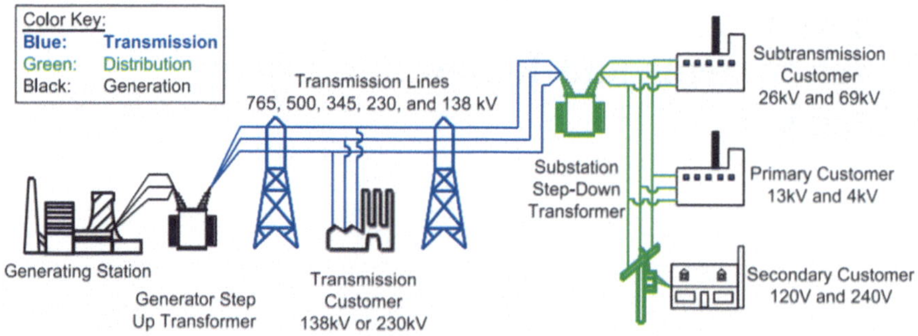


**Fig. 1.1** Basic structure of a power system [1]

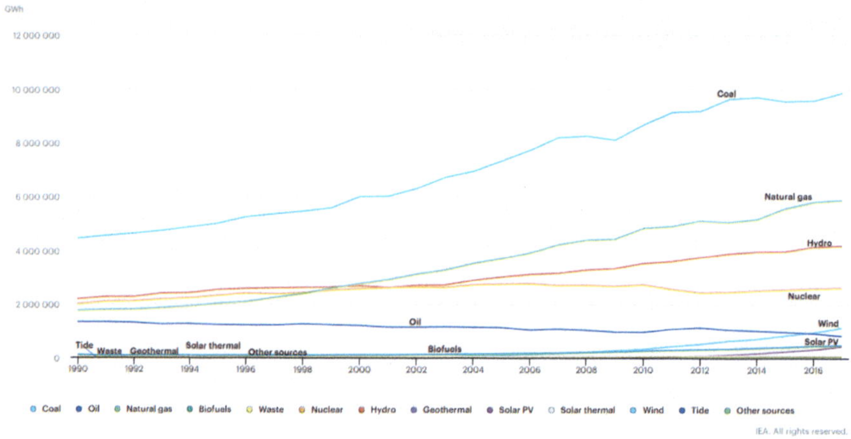


**Fig. 1.2** World electricity generation from RES and non-RES [2]

of source include solar, wind, biogas, and geothermal sources etc. Figure 1.2 shows the electricity generation of different energy sources from 1990 to 2016, inclusive of RES and non-RES. Figure 1.3 shows the total world electricity consumption in the same period.

## 1.1.2 *Transmission of Electric Power*

The electric power obtained from generating plants is transferred to distribution points in the power system. The network that is used for this long-distance transmission of power is called the transmission network. The transmission network is the interface between the generating stations and the distribution points in the power system. In view of the importance of transmission networks, these are often

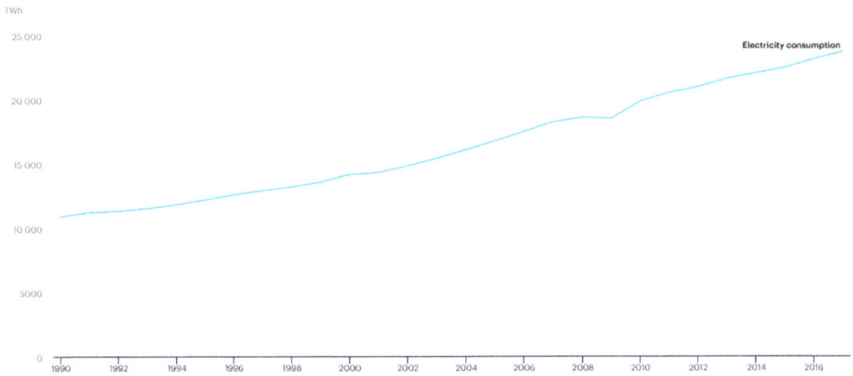


**Fig. 1.3** Total world electricity consumption [3]

built in such a way that multiple channels exist for power to flow from generating stations to loads in a bid to minimize disruptions. Because of the long distance most transmission networks span, the transmission network has to transmit power at high voltages in order to minimize losses in the system. Transmission networks are usually constructed to support three-phase high-voltage alternating current (HVAC) power. Special kinds of transmission lines are constructed to transmit direct current (DC). These are referred to as high-voltage DC (HVDC) transmission lines. HVDC lines are typically suitable to transfer electric power over very long distances. HVDC lines are efficient and yield lower power losses than other types. HVAC and HVDC transmission lines are usually deployed as overhead lines rather than underground. The choice of overhead construction lines is motivated by the fact that they are cheaper to maintain, even though they are more expensive to construct. Transmission lines' voltages are in the range of >66 kV with sub-transmission voltage levels in the range of 66–132 kV.

### *1.1.3 Distribution of Electric Power*

The distribution system is the last segment of the electric power supply process. Here electric power is conveyed from the transmission system to the end users. At the distribution level, voltage levels range from 2 to 33 kV. This voltage level is referred to as the distribution voltage level. These voltage levels are further stepped down to between 240 and 415 V, which is the range typically used by households and businesses. This is referred to as utilization voltage. Industrial consumers are often supplied at the distribution voltage level. The incorporation of generators at this spectrum of the power system, placing the generators closer to the end users, is referred to as distributed generation (DG). The incorporation of DG close to the consumer has a number of advantages, chiefly the fact that these generators improve the reliability and quality of supply. When these DG sources are renewable, they also yield significant environmental benefits.

Generators in the power system can therefore be incorporated in two key ways in the power system. The first is at the generation spectrum of the power system. These power sources are large and need to be transmitted across long distances in order to be used by consumers. On the other hand, DG enables the energy sources to be sited close to the consumer, thereby minimizing power losses and improving grid reliability. At both spectrums of the grid, RES can be used.

RES are sources of energy that can be readily replenished. Unlike non-RES, they have the added advantage of being environmentally friendly and sustainable. A brief description of prominent RES [4] is detailed below:

### *1.1.4 Hydropower*

Hydropower can be simply defined as power obtained from the kinetic energy of water. This power has historically been used to power mechanical machines such as mills. When energy stemming from fast-flowing water is used to produce electricity, it is referred to as hydroelectricity. Hydroelectricity is considered as renewable energy. Hydroelectricity does not release harmful emissions into the atmosphere even though it has environmental impacts. Hydropower is used in three principal ways in the electricity generation process. The predominant method is the use of a reservoir created by a dam, from which water is released and used to turn turbines and generate power. The second method is the "run of the river" approach [4], where water is steered away from a river into a turbine and electricity is generated directly. The final approach is the pumped hydro storage mechanism. In this approach, two reservoirs are created at different elevations. When the demand for electricity is low, water is pumped from the lower reservoir to the higher reservoir. The water in the higher reservoir is later released to generate power when the demand for electricity is high and more electricity needs to be generated. Figure 1.4 indicates the worldwide use of hydropower.

### *1.1.5 Solar Energy*

Solar energy entails the use of the sun's radiation to heat either space or water or to produce electricity. As a heating resource, solar energy can be either a passive or active heating source. As a passive heating source for homes, homes are constructed in such a way that sunlight can stream in during the day for space heating and floors are constructed with either dark tiles or bricks that will store heat in the daytime and diffuse it at night [4]. The homes are kept cool in summer by the dark bricks or tiles, which remain cool, and awnings are used to shade the windows from excessive sunlight. Solar energy as a passive heating source, coupled with energy-efficient

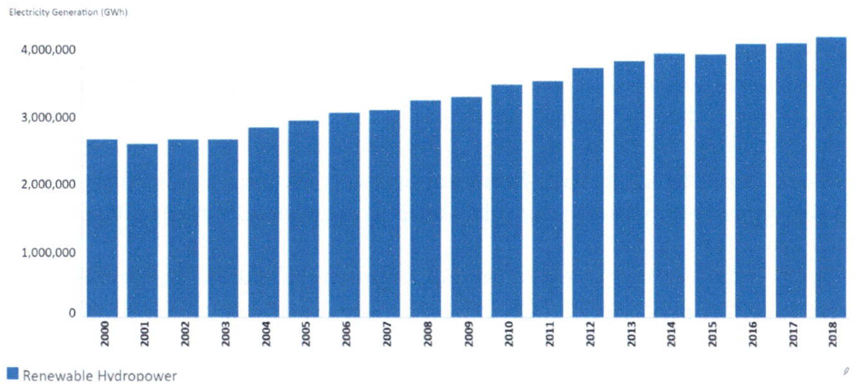


**Fig. 1.4** Worldwide electricity generation using hydropower (2000–2018) [5]

implementations in a home (energy-efficient appliances and building fixtures) have the potential to minimize energy consumption [4], which will yield a commensurate reduction in emissions.

The sun is also a good resource for water heating. Active solar heating works by constructing collectors, which are aligned to face the north and heats water either directly or through a heat transfer fluid. The water is then stored in a water tank and dispersed as in a typical water-heating system.

The quantity of hot water produced by a solar water heater is contingent on a number of factors, including the amount of solar radiation, the type of water-heating system, the size of the solar system, orientation and tilt angle of the collectors, among others.

There are essentially three types of solar water collectors [4], which define the types of solar water-heating systems. The first type, the most common type of solar water-heating collector, is the flat plate collector, which is made up of an impermeable transparent cover with a dark absorber plate underneath. The second type of collector is the evacuated tube collector, which is a collection of transparent glass tubes arranged in parallel. Each transparent glass tube is made up of the outer glass tube and an internal tube-like absorber. The internal tube is termed an absorber because it absorbs energy from the sun without corresponding heat loss. The third type of collector is the concentrating collector, which works by optically concentrating solar energy prior to its conversion to heat. This type of collector consists of a trough-like concentrator, which transfers direct beam radiation onto the receiver and a receiver component, which converts the absorbed solar radiation into some other form of energy.

The flat plate collector, evacuated tube collector and concentrated collector all heat water either as passive solar water-heating systems or active solar water-heating systems [4]. An active solar heating system circulates household water or heat transfer liquids through the solar collector to the water tank via an electric pump or other electric component. A passive solar heating system also circulates water to the tank from the collector, but does not use any electric component and achieves the water flow via gravity stemming from natural convection. Passive solar water-heating systems are generally more reliable owing to the absence of electric components, while active solar water-heating systems are generally more efficient, albeit expensive. A further distinction between the systems is that in colder climes, the passive water-heating systems tend to freeze, which does not happen to active water-heating systems [4]. Active water-heating systems are also typically hybridized with solar panels in the event of power failures so that water can still be heated.

Solar energy can also be used to produce electricity. The use of rays from the sun to generate electricity can be done in two principal ways, either via photovoltaic (PV) technology or solar thermal electric systems. PV solar systems contain semiconducting modules that convert solar radiation to produce DC electricity, inverters to convert the DC to alternating current and often battery storage systems to store excess generated electricity to be used at times when the generated electricity is insufficient to satisfy demand.

Auxiliary equipment in PV systems can include the PV hosting structure and protection equipment. The PV panels can be installed on top of buildings, i.e. rooftop power systems, and in general are efficient electricity providers.

In essence, PV solar systems convert the sun's radiation to electricity. Solar thermal electricity-generating systems convert the sun's radiation to heat, which is then used in turn to generate electricity. The most common solar thermal electricity-generating system is the concentrated solar power (CSP) system.

CSP generates electricity by using mirrors to "concentrate" the sun's rays onto a receiver, which uses the heat to power a conventional generator to generate electricity.

There are three types of CSP systems, namely the parabolic trough, power receivers, and dish Stirling systems [4]. The parabolic trough system uses parabolic-like mirrors to concentrate the sun's energy onto an oil-filled receiver, which runs along the trough's axis. When the oil is sufficiently hot, it is used to boil water that is deployed in a conventional steam generator for electricity generation.

The power receivers consist of a large field of mirrors, which concentrate the sun's rays onto the top of a tower where the receiver is sited. The generated heat is used to heat molten salt that flows through the receiver, which is in turn used to generate electricity via a conventional steam generator. The dish Stirling system uses a dish-shaped mirror whose surface concentrates the sun's heat onto a receiver sited at the dish's focal point. The generated heat from the receiver is transferred to a Stirling engine, which is then able to generate electricity [4].

PV systems can produce electricity on either a small or large scale, while CSP systems (solar thermal systems) typically produce electricity on a large scale.

Solar electricity-generating systems, especially when they form part of the RES in a microgrid, can be connected in the off-grid mode, grid-connected mode, or hybrid mode. The off-grid mode is one in which the RES only supplies power and there is no connection from the main grid. When there is grid supply in addition to the RES, it is referred to as grid-connected mode, while the hybrid mode is one with an off-grid solar system coupled with utility power or grid-connected solar power with backup battery storage (Fig. 1.5).

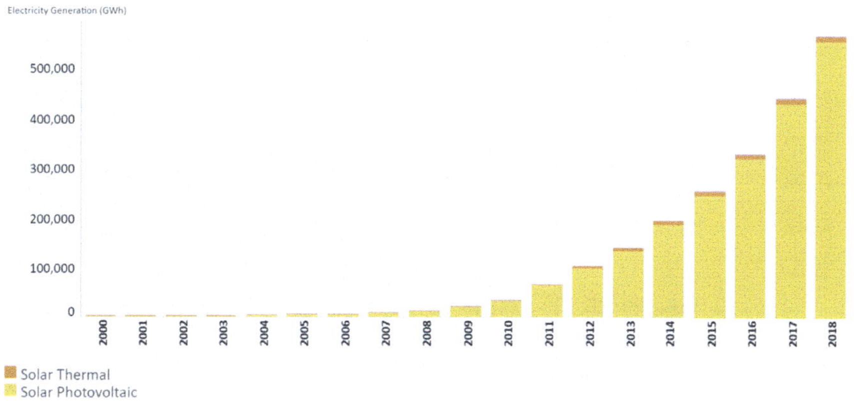


**Fig. 1.5** Worldwide electricity generation using solar energy (2000–2018) [5]

### *1.1.6 Wind Energy*

Wind electricity generation uses wind as the prime mover of wind turbines to generate electricity.

Wind turbines are electric power generators that convert the wind's kinetic energy to electrical power. They can rotate along either a vertical or horizontal axis. Horizontal axis wind turbines are more commonly found than vertical axis wind turbines. The wind turbine blades spin around a central hub, which is then connected to a shaft that is the prime mover of a mechanical drive system using an electric generator to generate electricity.

A collection of wind turbines is called a wind farm and this can be spread across a large expanse of land. Wind farms can be sited either onshore or offshore. Onshore wind farms are cheaper to construct than offshore wind farms [6]; however, because wind speeds are stronger offshore, offshore wind farms can potentially generate more electricity.

Wind turbines, just like PV systems, can be used to generate electricity for small-scale and large-scale uses and can be connected to the grid (Fig. 1.6).

### *1.1.7 Bioenergy*

Bioenergy is energy obtained from biomass. Biomass is organic matter and can originate from plants, animals, or waste products. Bio-power entails generation of power from biomass. This can be done by converting biomass to biogas, which produces methane; the biogas is then used as fuel in a gas turbine and electricity is generated. The conversion of biomass to biogas typically produces methane and carbon dioxide ($CO_2$) as gases. Biomass can also be converted to liquid, which can serve as a fuel. This is often referred to as biofuel [4]. The most popular biofuel is ethanol,

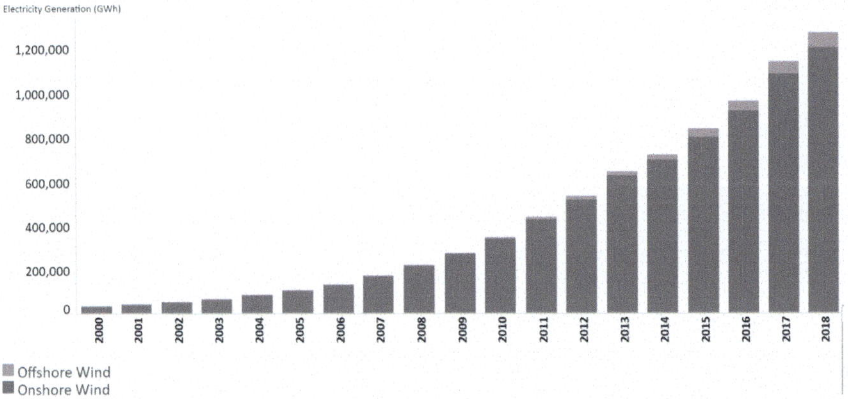


**Fig. 1.6** Worldwide electricity generation using wind (2000–2018) [5]

which is obtained by fermenting biomass with a high carbohydrate content. Biofuels have high energy density and have increasingly found industrial application [4].

Biomass can also be converted to pyrolysis oil [4], which can be used for power generation, among other uses. Pyrolysis is obtained from the heating of biomass in a reactor at very high temperatures in the absence of oxygen. The oil that is obtained is then cooled. Another liquid fuel to which biomass can be converted is called biodiesel, which is obtained from animal and vegetable fats. This biofuel has also found industrial application as either a vehicle fuel or fuel additive, as it emits fewer greenhouse gases than conventional vehicle fuels. A further use of biomass in the power-generation process is its integration with coal in coal-fired power stations. Incorporating biomass into the coal feed stock reduces the harmful emission of greenhouse gases, which reduces the environmental impact of burning coal and leads to better air quality [4] (Fig. 1.7).

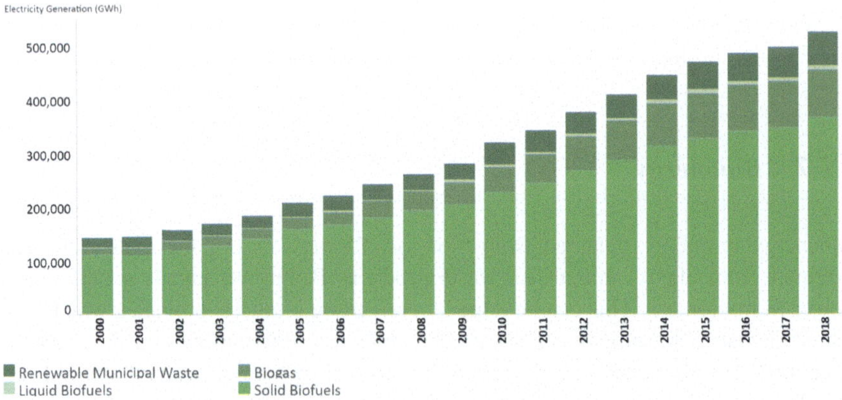


**Fig. 1.7** Worldwide electricity generation from bioenergy (2000–2018) [5]

### 1.1.8 Geothermal Energy

Geothermal energy is energy hidden in the earth's core. Temperatures in the earth's core can be as high as 4982.2 °C and the heat from this core forms pools or reservoirs of water and steam [4]. This heat can then be leveraged for electricity generation and used to satisfy heating and cooling demands.

For electricity generation, geothermal power plants are used. These power plants work by obtaining access to the underground steam/water in the earth's core via a well and using this to power a conventional steam/water turbine, thus generating electricity [4]. The water is usually returned to the earth's core, replenishing the reservoir and mitigating losses. There are three principal types of geothermal plants, depending on the type of reservoirs from which they draw [4]. The first is called a dry steam plant, which produces electricity by drawing on dry steam from the earth's core. The second is called a flash steam plant and a third binary cycle plant. Both flash steam and binary cycle plants use hot water reservoirs to generate electricity. While binary cycle plants can operate using water at temperatures between 107.2 °C and 182.2 °C, flash steam plants can operate at water temperatures higher than 182.2 °C [4].

As stated earlier, geothermal energy can also be used for heating and cooling applications. The first occurs by using geothermal heat pumps where the stable temperature of shallow ground (between 10 °C and 21.1 °C), is used for heating buildings in summer and cooling buildings in winter [4]. Geothermal energy can also be used as a "direct use" energy source [4]. In this system, the hot water from the water reservoir is passed through a heat exchanger, which also has a working fluid. The working fluid is kept separate from the hot water but is heated by the hot water. The working fluid is then passed through pipes and is used directly as a heat source. The cooled water is often disposed of in a disposal system. Direct use of geothermal energy is primarily made to heat buildings and it can also be used for agricultural purposes. The heating can be done on a large scale and supply many buildings, which is why it is often termed district heating. Direct use systems are also applied to melt snow on roads and sidewalks [4] (Fig. 1.8).

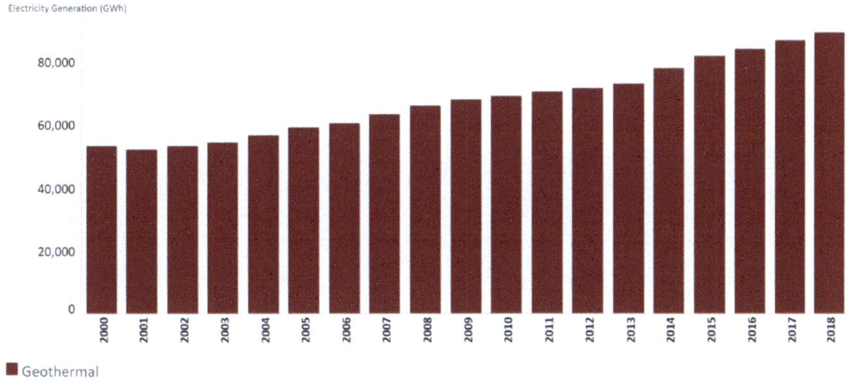


**Fig. 1.8** Worldwide electricity generation using geothermal energy (2000–2018) [5]

### *1.1.9 Ocean Energy*

The ocean can be used to produce electricity in four principal ways [7], namely ocean thermal energy, tidal energy, wave energy and salinity gradient energy. Ocean thermal energy is produced from the sun's heat; the surface water of the ocean is heated and used to turn a turbine, which in turn generates electricity. Tidal energy uses a dam to move water through turbines, which is then used to generate electricity. Wave energy is the utilization of energy in waves to propel a generator via a turbine or using the height difference in waves to propel a turbine.

Salinity gradient energy is electricity generation based on the difference in salt concentration between two bodies of water. Pressure-retarded osmosis processes utilize this difference to generate electricity.

It is worth noting that ocean energy technology is not as advanced as other technologies and hardly finds deployment (Fig. 1.9).

**Hydrogen**

Hydrogen is one of the universe's most abundant elements, but is not a naturally occurring gas [4]. This therefore means hydrogen can only be produced by splitting it from compounds. There are several methods of producing hydrogen, including the electrolysis of water, gasification, and reformation of natural gas or the partial oxidation of methane. The most common method of producing hydrogen is known as reforming [4]. This entails the use of heat to split hydrogen molecules from carbon. This splitting process requires energy and leads to emissions of $CO_2$. If this process is powered by RES without emissions, the hydrogen is referred to as $CO_2$ neutral hydrogen or green hydrogen.

Hydrogen is a key ingredient in the operation of fuel cells. Fuel cells, via redox reactions, convert the energy from hydrogen into electricity using an oxidizing agent (oxygen). Unlike conventional batteries, fuel cells need a constant stream of

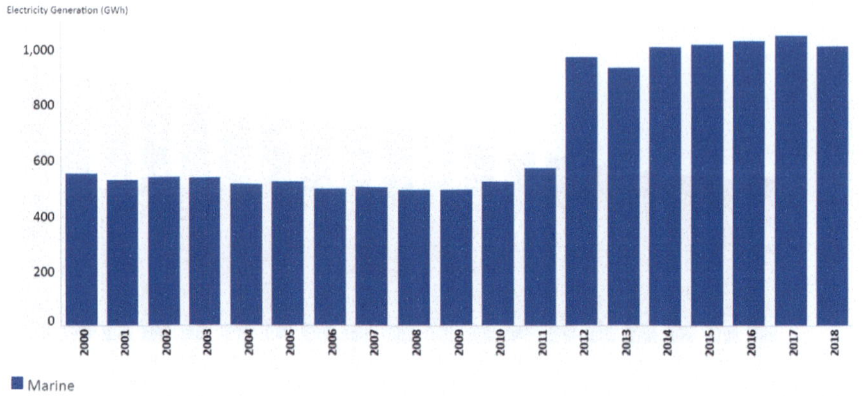


**Fig. 1.9** Worldwide electricity generation using ocean/marine energy (2000–2018) [5]

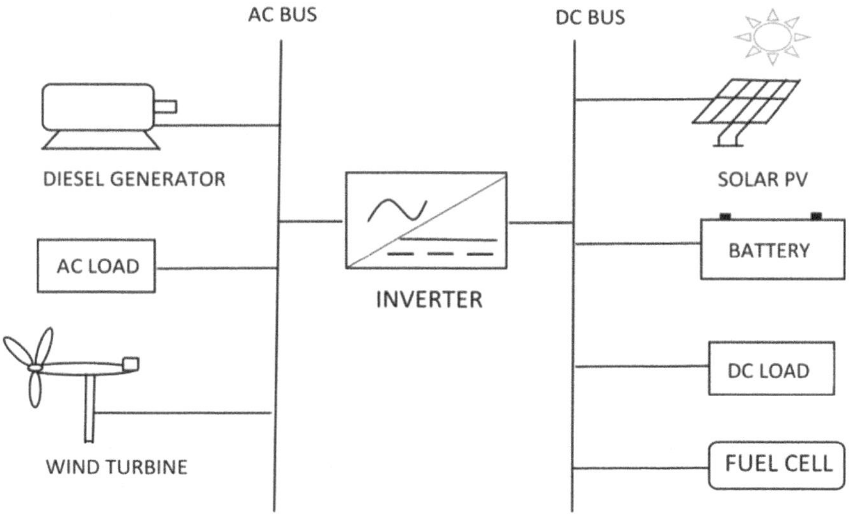


**Fig. 1.10** Configuration of RES integrated into a microgrid [8]

fuel (hydrogen) and oxygen to sustain chemical reactions and by extension operations. They can thus produce electricity continuously, as long as these two elements are in continuous supply. Hydrogen fuel cells are used in both low-power and high-power applications, among others in automobiles, transportation, and food processing [4].

Figure 1.10 below shows an example of a microgrid with a number of RES as the power source and battery storage.

## 1.2 Smart Grids

A smart grid means different things to different professionals and organizations; therefore, it is difficult to agree on a common or general technical definition. What can, however, be agreed upon is that a smart grid leads to the operational improvement and reliability of the power grid. The following are definitions of smart grids from various sources:

The Energy Independence and Security Act of 2007 (EISA 2007) defines a smart grid as [9]

> the modernization of the Nation's electricity transmission and distribution system to maintain a reliable and secure electricity infrastructure that can meet future demand growth.

The International Energy Agency (IEA) defines a smart grid as [10]

> an electricity network that uses digital and other advanced technologies to monitor and manage the transport of electricity from all generation sources to meet the varying

electricity demands of end users. Smart grids co-ordinate the needs and capabilities of all generators, grid operators, end users and electricity market stakeholders to operate all parts of the system as efficiently as possible, minimising costs and environmental impacts while maximising system reliability, resilience and stability.

The International Electro-technical Commission (IEC) defines a smart grid as [11]

> an electricity network that can intelligently integrate the actions of all users connected to it – generators, consumers and those that do both – in order to efficiently deliver sustainable, economic and secure electricity supplies. A smart grid employs innovative products and services together with intelligent monitoring, control, communication, and self-healing technologies to: facilitate the connection and operation of generators of all sizes and technologies; allow consumers to play a part in optimizing the operation of the system; provide consumers with greater information and choice of supply; significantly reduce the environmental impact of the whole electricity supply system; deliver enhanced levels of reliability and security of supply.

The United States Office of Electricity Delivery and Energy Reliability [12] defines a smart grid as a.

> class of technology that people are using to bring utility electricity delivery systems into the twenty-first century, using computer-based remote control and automation. These systems are made possible by two-way communication technology and computer processing that has been used for decades in other industries. They are beginning to be used on electricity networks, from the power plants and wind farms all the way to the consumers of electricity in homes and businesses. They offer many benefits to utilities and consumers – mostly seen in big improvements in energy efficiency on the electricity grid and in the energy users' homes and offices.

Furthermore, the Energy Independence and Security Act of 2007 details ten distinguishing features of a smart grid [9], reproduced verbatim:

(1) Increased use of digital information and control technology to improve reliability, security, and efficiency of the electric grid;
(2) Dynamic optimization of grid operations and resources with full cybersecurity;
(3) Deployment and integration of distributed resources and generation, including renewable resources;
(4) Development and incorporation of demand response (DR), demand-side resources, and energy efficiency resources;
(5) Deployment of "smart" technologies (real-time, automated, interactive technologies that optimize the physical operation of appliances and consumer devices) for metering, communications concerning grid operations and status, and distribution automation;
(6) Integration of "smart" appliances and consumer devices;
(7) Deployment and integration of advanced electricity storage and peak-shaving technologies, including plug-in electric and hybrid electric vehicles as well as thermal-storage air conditioning;
(8) Providing consumers with timely information and control options;
(9) Development of standards for communication and interoperability of appliances and equipment connected to the electric grid, including the infrastructure serving the grid;

(10) Identification and lowering of unreasonable or unnecessary barriers to adoption of smart grid technologies, practices, and services.

The core philosophy behind a smart grid is to transmit energy seamlessly from generators (both conventional and renewable powered generators) to end users in a cost-effective, reliable manner with advanced and intelligent infrastructure that permits the monitoring and control of network components and power flow. It is worth noting that in the context of a smart grid, the end users are not referred to solely as consumers but as prosumers [13, 14] because they simultaneously produce and consume electrical power. Monitoring and control of the power system components should be done in real time with the possibility of prompt remedial action in the event of blackouts or brownouts. The smart grid is also contingent on the implementation of advanced information communication infrastructure and digital technologies for effective condition monitoring and enhanced reliability. These technologies are applied at different spectrums of the power grids.

Examples of such technologies include smart meters, flexible alternating current transmission system devices (FACTS) devices, vehicle-to-grid technologies, supervisory control and data acquisition, grid-to-vehicle technologies, and phasor measurement units (PMU) among others.

The net result of these technologies is near real-time power system observation and control and the amplified incorporation of DR programs. DR programs seek to incentivize electricity consumers to curtail their electricity demand in response to price signals and customer infrastructure required [15, 16]. Other key components of the rapidly expanding smart grid include various energy and heat storage devices, the electrification of the transport system, prosumers, electric vehicles and RES [17]. Figure 1.11 shows a snapshot of the smart grid.

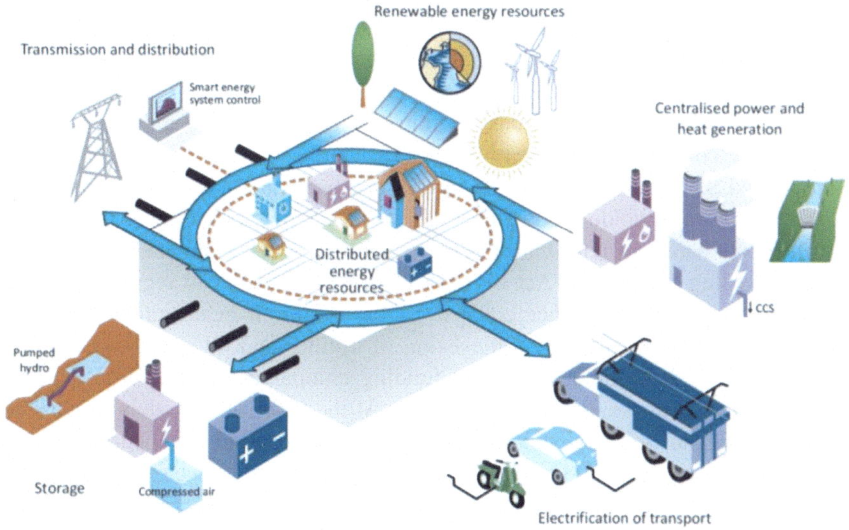


**Fig. 1.11** Overview of a smart grid [18]

## 1.3 Electricity Deregulation

The electricity industry is broadly broken down, based on its functions, into generation, transmission, and distribution. Generation is the creation of electrical power or electricity that occurs via the conversion of mechanical energy to electrical energy. The electric power obtained from generating plants is then transferred to distribution points in the power system. The distribution system is the last portion of the electric power supply process. Here electric power is transferred from the transmission system to the end users. Electric utilities use various mechanisms to perform these three core functions. The traditional mechanism adopted by electric utilities is referred to as vertical integration. Vertical integration occurs when a utility entity is solely in charge of generation, transmission, and distribution. Vertical integration has a host of advantages in theory, chief of which is the fact that having one entity control the whole spectrum of the power grid will lead to increased efficiency and minimal operating costs. A single entity controlling the whole spectrum of the grid from generation to distribution means that it could make the required capital-intensive investments in the electrical grid infrastructure and be poised to recoup investments with fair pricing. This pricing strategy for electric power is known as cost-based pricing and has been used by electric utility monopolies across the world.

Vertical integration led to the birth of electricity utility monopolies. At the beginning of the electricity industry in the United States (prior to 1930), these monopolies were unregulated and were referred to as holding companies [19]. Holding companies were business entities formed by financiers and served as a vessel for a diverse range of business. Electricity production was just a business within the package and was generally profitable. Holding companies use the income from the sale of electricity to finance relatively riskier business ventures in the grouping. This led to a scenario where these holding companies were financially exposed, as they had little or no assets, only huge financial obligations and massive debts.

The Great Depression had a massive impact on the electricity industry. At that time, holding companies with their business structure accounted for a large portion of the United States' electricity production. When banks needed money and sought to recover this from holding companies, these companies folded and left many of their investors bankrupt. This catastrophic event led to the enactment of the Public Utilities Holding Companies Act of 1935 [19], which sought to regulate holding companies and prevent another collapse of the sector. It achieved this by prohibiting holding companies from consisting of utilities and other businesses. The Act also restricted the amount of debt holding companies could accrue and the geographical areas these utilities could serve. At the same time, Federal organs of state (such as the Security and Exchange Commission and the Federal Power Commission, now Federal Energy Regulatory Commission) were charged with powers to regulate interstate electricity transmission, while states had the prerequisite power to regulate electricity transmission within their state borders [19].

States could also determine and set the electricity retail price in their domains (Brennan et al. 1996), which factored in the utility's cost of power production and marginal profit (cost-based pricing). If the utilities wanted a price increase, they had to show that there was a corresponding price increase in production costs and this had to be approved by the relevant regulator.

However, advances in generation technologies led to a mounting push to introduce competition in the generating spectrum of the grid. The rationale for this is that competition in this segment will propel firms to utilize optimal energy-producing technologies at minimal cost. This push was also assisted by the drive to do away with fossil fuel-powered generator sources with their attendant environmental implications. Thus, industry players were being put under pressure to introduce a market at the generating spectrum of the grid. There was also a drive for introducing competition at the distribution spectrum of the grid. It was believed that having firms compete for customers at this spectrum of the grid would lead to the consumer being able to choose the cheapest source of electricity. This was to combat the ability of electric utilities (obtainable from vertical integration) to fix electricity prices arbitrarily; such tariffs were then passed on to the electricity consumers. Competition was envisaged to create efficiency in the distribution of electricity. The transmission spectrum of the grid historically was spared the drive for competition. This was in acknowledgement of the fact that there were (national) security implications of creating a market out of this spectrum.

The introduction of competition in the generation and demand spectrum of the power grid led to the incorporation of market pricing in the pricing of electricity instead of cost-based pricing. Market pricing meant electricity was priced based on the interplay between forces of demand and supply.

It is noteworthy that deregulation of the electricity industry was adopted for various reasons depending on whether it happened in a developed nation or a developing nation. In developed nations, deregulation was spurred on by the conventional reasons of price reduction, a drive for efficiency and minimizing the involvement of government. In developing nations/economies, electricity sector deregulation was motivated by the need to attract investors in the electricity sector in order to improve power system operations and reliability. However, the overarching aim (in either developing or developed nations) of deregulation was to improve efficiency and minimize the price of electricity, though this meant different things in different contexts. This ambiguity has also made it difficult to assess different deregulation market structures and the degree of their success uniformly [20].

The drive for deregulation of the electricity industry has also not had a uniform start in various jurisdictions. In the 1980s Chile led the movement with the enactment of the 1982 Electricity Act. England and Wales enacted the Electricity Act in 1989. Norway adopted its new Energy Act in 1990 and Colombia embraced deregulation in 1995, whereas Nigeria enacted the Electric Power Sector Reform Act in 2005. Various European nations followed the lead of Norway and by 2000, the European Union had mandated deregulation among its member nations [19]. In the United States, deregulation picked up steam in the 1980s in response to increasing electricity prices. By 1992, The Energy Policy Act had been enacted and paved the way for the deregulation of the electricity industry by states. California and Pennsylvania were among the first states to adopt deregulation and move from cost-based pricing to market pricing.

Moving forward, 23 states had enacted laws to advance electricity deregulation by 2000, although the electricity crisis in California in that year dampened enthusiasm for deregulation. By 2014, seven states had abandoned their drive for deregulation, 16 states had active electricity markets and the remainder did not participate in deregulation [19] (Table 1.1).

**Table 1.1** Overview of electricity deregulation in organization for economic cooperation and development countries [20]

| Country | Liberalization | Electricity market | Consumption (TWh), 2002 |
|---|---|---|---|
| Australia | Electricity Industry Act for Victoria (1994) | National Electricity Market (1997), Victoria Pool (1994) | 196 |
| Austria | Law of Electricity Supply (1998) | Energy Exchange Austria (2002) | 55 |
| Belgium | Law for the Organization of the Electricity Market, 29 April 1999 | None | 79 |
| Canada | Alberta—Electric Utilities Act (2001), Ontario—Energy Competition Act (1998) | Alberta Pool (1996) Ontario Market (2002) | 487 |
| Czech Republic | The Energy Act (2000) | Opera´tor trhu s elektrinou (2002) | 55 |
| Denmark | Amendment to Danish Supply Act (1996, implemented 1998) | Nord Pool (1999 East Denmark; 2000 West Denmark) | 32 |
| Finland | Electricity Market Act (1995) | Finnish Electricity Exchange (1995) | 79 |
| France | Law No. 2000–108 (2000) | Powernext (2001) | 415 |
| Germany | Act on the Supply of Electricity and Gas (1998) | European Electricity Exchange (2000) | 513 |
| Greece | Electricity Law, 21 December, 1999 | None | 47 |
| Hungary | Electric Power Act (2001) | None | 36 |
| Iceland | Electricity Act (2003) | None | 8 |
| Ireland | Electricity Regulation Act (1999) | Trading and Settlement Market (2000) | 22 |
| Italy | Bersani Decree (1999) | Electricity Market (2004) | 294 |
| Japan | Amendments to Electric Utility Law (1995) | None | 971 |
| Korea | Act on Promotion of Restructuring of the Electric Power Industry (2000) | Korea Power Exchange (2001) | 267 |

(continued)

**Table 1.1** (continued)

| Country | Liberalization | Electricity market | Consumption (TWh), 2002 |
|---|---|---|---|
| Luxembourg | Law of 24 July 2000 on the organization of the electricity market | None—most power imported from neighbors | 6 |
| Mexico | Independent power producers allowed, no further liberalization | None | 190 |
| Netherlands | The Electricity Act (1998) | Amsterdam Power Exchange (1999) | 101 |
| New Zealand | Energy Act and Companies Act (1992) | Electricity Market Company (1996) | 36 |
| Norway | Energy Act (1990) | Norwegian Power Pool (1991), Nord Pool (1996) | 107 |
| Poland | Energy Act 1997 | Polish Energy Exchange (2000) | 117 |
| Portugal | Decree Laws 182/95, 183/95, 184/95, and 185/95 of 27 July 1995 | Iberian electricity market (2004) | 42 |
| Slovakia | Law on Energy (1998) | | 24 |
| Spain | Electricity Act (1994) | OMEL (1997), Iberian electricity market (2004) | 218 |
| Sweden | Law for the Supply of Electricity 10/95 | Nordpool (1996) | 138 |
| Switzerland | Act defeated in 2002 referendum | None | 55 |
| Turkey | Electricity Market Law (2001) | | 118 |
| United Kingdom | Electricity Act (1989), | Electricity Pool of England and Wales (1990), NETA (2001) | 344 |
| United States | PURPA (1978), Energy Policy Act (1992) | PJM (1998), ISO-NE (1999), NewYork (1999), ERCOT (2002) | 3660 |

## 1.4 Operation and Control

The operation and control of a power system is concerned with the optimal management of the power system inclusive of its power sources, transmission lines and load. This very difficult task requires highly skilled personnel, operations, and algorithms.

The North American Electric Reliability Council has delineated key areas for optimal operation and control of power systems [1]:

i. Power generation and demand must be balanced.
ii. Reactive power supply and demand must be balanced to maintain acceptable voltage levels.
iii. Transmission line power flows must be monitored constantly to avoid line limits violations.
iv. It must be ensured that power system stability and reliability are maintained.
v. Preparation for emergencies must be such that reliability is maintained even if an important component fails.

This includes conventional power system operational and control issues such as power flow, optimal power flow (OPF), unit commitment, economic dispatch, generator maintenance scheduling (GMS), generator and transmission expansion planning, FACTS devices placement, and PMU placement, etc. These issues have various time scales ranging from seconds (optimal power flow) to weeks (generator maintenance scheduling) and years (generator and transmission expansion planning).

Mathematical optimization can be a useful tool in achieving optimal operation and control of the power system for the aforementioned power system issues. A mathematical model typically maximizes or minimizes power system parameters. This can be the minimization of generator fuel costs, the maximization of output of power generators, minimization of generator emissions, and minimization of operating costs, etc.

Recent advances in current power systems have introduced modern operational issues such as the grid integration of renewables, DR, prosumer integration, and electric vehicles, etc.

The advent of the Fourth Industrial Revolution and its associated technologies such as the internet of things (IoT), blockchain, machine learning, big data etc. have also been deployed to increased grid reliability and stability. IoT and blockchain technologies have been used for smart home management systems and have led to more efficient operation and control of the power system. Classification, clustering, and regression have been used for solar radiation modeling, wind speed modeling, price forecasting, and load forecasting, etc. [21].

## 1.5 Summary

This chapter introduced a power system and detailed the various sectors in power system operation. Furthermore, recent RES that have increasingly been adopted were discussed. The chapter also examined various definitions of a smart grid and what constitutes a smart grid. The deregulation of the electric power industry across various climes was mentioned and the chapter was concluded with an introduction to operational and control issues in a power system.

## References


1. U.S.-Canada Power System Outage Task Force (April 2004) "Final report on the August 14, 2003 blackout in the United States and Canada: causes and recommendations"
2. IEA, "Data and Statistics—IEA" (2020) [Online]. Available: https://www.iea.org/data-and-statistics?country=WORLD&fuel=Energy%20supply&indicator=Electricity%20generation%20by%20source. [Accessed: 09 Jul 2020]
3. IEA, "Data and Statistics—IEA" (2020) [Online]. Available: https://www.iea.org/data-and-statistics?country=WORLD&fuel=Energy%20consumption&indicator=Electricity%20consumption. [Accessed: 09 Jul 2020]
4. Tromly K (2001) "Renewable energy: an overview," National renewable energy, laboratory, golden, CO (US)
5. International Renewable Energy Agency (2020) "Trends in Renewable Energy" [Online]. Available: https://public.tableau.com/views/IRENARETimeSeries/Charts?:embed=y&:showVizHome=no&publish=yes&:toolbar=no. [Accessed: 09 Jul 2020].
6. Blanco MI (2009) The economics of wind energy. Renew Sustain Energ Rev 13(6–7):1372–1382
7. García Vera Y, Dufo-López R, Bernal-Agustín J (2019) Energy management in microgrids with renewable energy sources: a literature review. Appl Sci 9(18):3854
8. Irena.org (2020) "Ocean energy" [Online]. Available: https://www.irena.org/ocean. [Accessed: 09 Jul 2020]
9. Energy Independence and Security Act of 2007
10. IEA (2011) Technology roadmap—smart grids. Paris: IEA. [Online]. Available: https://www.iea.org/reports/technology-roadmap-smart-grids. [Accessed: 09 Jul 2020]
11. Iec.ch (2020) "IEC—Smart grid > Background—What is Smart Grid?" [Online]. Available: https://www.iec.ch/smartgrid/background/explained.htm. [Accessed: 09 Jul 2020]
12. Energy.gov (2020) "Grid modernization and the smart grid" [Online]. Available: https://www.energy.gov/oe/activities/technology-development/grid-modernization-and-smart-grid. [Accessed: 09 Jul 2020]
13. Damisa U, Nwulu NI, Sun Y (2018) A robust energy and reserve dispatch model for prosumer microgrids incorporating demand response aggregators. 10(5)
14. Damisa U, Nwulu NI, Sun Y (2018) Microgrid energy and reserve management incorporating prosumer behind-the-meter resources. IET Renew Power Gener 12(8):910–919
15. Fahrioglu M, Nwulu NI (2012) Investigating a ranking of loads in avoiding potential power system outages. J Electr Rev (Przeglad Elektrotechniczny) 88(11a):239–242
16. Nwulu NI, Fahrioglu M (8–11 May 2011) "A neural network model for optimal demand management contract design", Proceedings of the 10th international conference on environmental and electrical engineering (EEEIC2011), Rome, Italy
17. Gbadamosi S, Nwulu NI, Sun Y (2018) Multi-objective optimization for composite generation and transmission expansion planning considering offshore wind power and feed in tariffs. IET Renew Power Gener 12(14):1687–1697
18. IEA (2014) Energy technology perspectives 2014, IEA, Paris https://www.iea.org/reports/energy-technology-perspectives-2014. [Accessed: 09 Jul 2020]
19. Owens S (2017) "Measuring the effect of electric utility deregulation on residential retail prices in a Midwestern state". MPA/MPP Capstone projects, University of Kentucky
20. Alsunaidy A, Green R (2006) Electricity deregulation in OECD (organization for economic cooperation and development) countries. Energy 31(6–7):769–787
21. Nwulu NI, Agboola PO (18–25 May 2012) "Modelling and predicting electricity consumption using artificial neural networks", Proceedings of the 11th international conference on environmental and electrical engineering (EEEIC2012), Venice, Italy
22. Unece.org (2020) [Online]. Available: https://www.unece.org/fileadmin/DAM/energy/se/pdfs/eneff/eneff_h.news/Smart.Grids.Overview.pdf. [Accessed: 09 Jul 2020]

# Chapter 2
# Demand Side Management

## 2.1 Introduction

The advancement in the power system and its rapid modernization is often referred to as the smart grid. Today's smart grid is grappling with increased demand due to a rapidly increasing population and increased demand for operational reliability. Power system regulators and operators have two principal ways of dealing with these challenges. The primary method is increasing the capacity of the power system, from the generation spectrum through the transmission spectrum to the distribution spectrum. The problem with this approach is that it often involves a huge financial outlay to which the utility and consumers might not be favorably disposed, as these expenses are usually passed on to the consumers. An allied reason is the fact that there are often environmental implications to expanding the capacity of the power system. It is widely known that power system plants, transmission lines and infrastructure generally leave a huge environmental footprint and depending on some technologies, might contribute to increased emissions and global warming.

The second approach, which is favorably regarded by power system operators and regulators, involves DSM programs. DSM programs are conceptualized to encourage consumers to be participants in the power system by curtailing their power consumption through a host of programs, behavioral changes and technological inventions.

One of the earliest advocates for DSM programs was Amory Lovins [1] who posited that "A kilowatt-hour saved is just like a kilowatt-hour generated … so they should be treated alike." Amory Lovins, in "Saving Gigabucks with Negawatts" [2], defined a negawatt as the unit of electricity saved and the inverse of the watt, which is the unit of energy consumed.

In essence, the aim of DSM programs is to alter a consumer's electricity consumption pattern actively and reduce energy consumption. This aim can be attained by peak clipping, valley filling, load shifting, strategic conservation, strategic load



N. Nwulu and S. L. Gbadamosi, *Optimal Operation and Control of Power Systems Using an Algebraic Modelling Language*, Green Energy and Technology,
https://doi.org/10.1007/978-3-030-00395-1_2

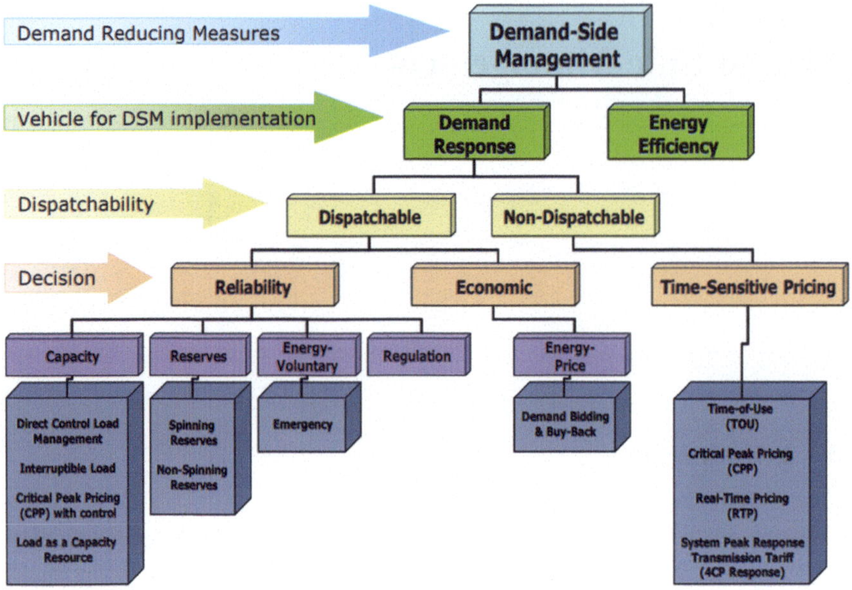


**Fig. 2.1** An overview of DSM, energy efficiency and demand response programs [3]

growth and flexible load shape [3]. DSM programs can be applied as either short-term or long-term strategic interventions that can be deployed in times of power system strain and stress. Furthermore, they can be deployed either throughout the power system or at previously identified problem points. DSM can be categorized into two, namely energy efficiency programs and demand response programs. Both programs are briefly detailed in the next subsections. Figure 2.1 presents a detailed overview of DSM programs.

## 2.2 Energy Efficiency Programs

Energy efficiency programs seek to curtail consumers' electricity consumption while maintaining the consumers' operational abilities. These programs are conceptualized to reduce the energy consumed by buildings, organizations or communities without altering the activities for which the consumers use the energy. Energy efficiency programs are independent of the location of the end users and the time profile of energy consumption. An example is replacing incandescent bulbs with compact fluorescent tubes, which will reduce the energy consumed by the user independent of the location of the user and the lighting time profile of the user. Energy efficiency programs are a significant cost-saving measure for consumers and utility alike and lead to a significant reduction in environmental emissions. Energy efficiency programs/investments also yield return on investment owing to the cost savings the consumer or utility accrues. To contextualize or explain energy efficiency programs properly, we will detail an example of an energy efficiency program below.

### *2.2.1 Building Retrofit Energy Efficiency Programs*

When building retrofit programs are undertaken, consumers replace their inefficient energy devices with energy-efficient ones. These programs are a classical straightforward kind of energy efficiency program and have the potential to save costs (especially for large buildings and other structures) and to minimize emissions.

To provide more insight into the structure and objectives of energy efficiency programs, we will detail a building retrofit program in South Africa, called the Standard Product Program [4].

The Standard Product program was designed by Eskom, the South African utility monopoly, to encourage various kinds of customers (commercial, industrial and agricultural customers) to discard their existing energy-inefficient devices or technologies and substitute those with energy-efficient fixtures or technologies. These replacement technologies or fixtures can range from a single component to an energy-saving system and may even be lighting fixtures, energy-efficient showerheads or heat pumps etc.

Eskom defined a framework to guide the choice of technologies to be contained within the Standard Product Program, which are that the technology should be [4]:

- A specific 'off-the-shelf' product, e.g. a lamp, fitting, heat pump, etc. satisfying a specific requirement, which appeals to a wide market;
- Approved by Eskom to be used as a replacement for less energy-efficient technologies (compact fluorescent lamps, incandescent bulbs, etc.);
- A component, sub-assembly or complete "machine"; and/or
- Programmable; it may even be a control system, provided a predicable saving can be attached to the technology.

Eskom will give participating customers rebates, since the customers would have incurred costs when installing the energy-efficient technologies.

Eskom has stipulated the following conditions:

- Total demand reduction must be between 1 and 250 kW.
- The total energy savings should exceed 2 MWh/a.
- The maximum rebate/technology is restricted to ZAR 750 000.
- The customer should dispose of all inefficient technologies and foot the bills for the installation of the new technologies.
- The customer should possess the appropriate tax clearance and their electricity account should not be in arrears.

## 2.3 Demand Response Programs

DR, one of the principal DSM techniques, can be simply defined as "a change in electric usage by end-use customers from their normal consumption patterns in response to changes in the price of electricity over time, or to incentive payments designed to induce lower electricity use at times of high wholesale market prices or when system reliability is jeopardized" [5, 6].

It follows that DR programs have the goal of reducing the consumer's electricity load profile but uses either incentives or time-varying price signals in a bid to achieve this aim. DR programs have the added advantage of minimizing operating costs, minimizing emissions and enhancing the reliability and efficiency of power systems. These advantages hold true for both electricity markets and vertically integrated power systems. However, in electricity markets, DR programs lead to reduced wholesale market prices [7].

There are two broad-based classifications of DR programs. The first type of DR program is price-based demand response [6]. In price-based DR programs, there is a time variation in electricity price [6]. This means that the same quantity of electricity will be priced differently, contingent on the time of day. The pricing structure largely depends on the existing demand, as demand varies depending on the time of day. Thus, when demand for electricity is high, electricity is expensive and when demand is low, electricity is priced lower. This has the advantage of altering consumers' behavior and causing them to use electricity at times when the price is lower. Examples of price-based DR programs include time-of-use rates (TOU), real-time pricing (RTP), critical peak pricing (CPP), extreme day pricing, and extreme day—critical peak pricing [6, 8].

The second type of DR program is IB-DR programs [9]. IB-DR programs, as the name implies, offer incentives to consumers who curtail their electrical load at times of power system pressure. The incentive can take the form of either lower electricity tariffs or monetary payments. IB-DR programs often include penalties in the event that consumers do not curtail their electricity at the required time and to the required degree. Examples of IB-DR include direct load control (DLC), interruptible services (IS), emergency demand response programs (EDRP), capacity market programs (CMP), demand-bidding/buyback programs and ancillary market services [9, 10]. Consumers are not compelled to participate in DLC and EDRP and there are no penalties. The two programs with penalties for defaulting consumers are IS and CMP.

Figure 2.2 shows the different types of DR programs.

### *2.3.1 Price-Based Demand Response*

Historically, tariffs for electricity consumers have been fixed and set in advance by utilities and regulators [11]. As the electricity industry experienced a transition from

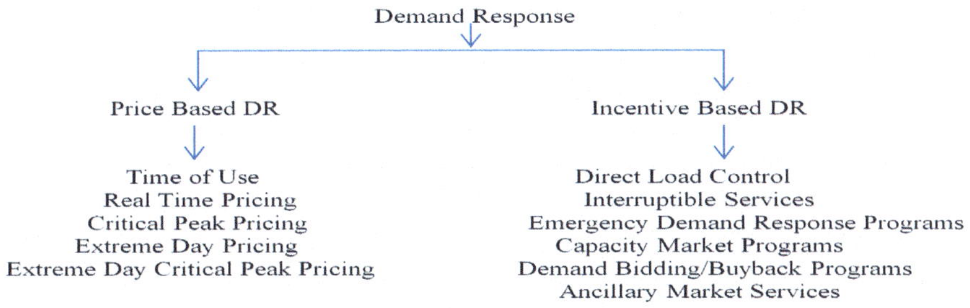


**Fig. 2.2** The classification of DR programs [8]

vertical integration to a market, there was a corresponding drive for pricing schemes that mirrored the price variation of the power system, since power system operational costs vary daily, weekly or seasonally. The expectation is that customers would leverage the price variation and adjust their consumption to take advantage of cheaper prices [6, 10]. In the next subsections, we briefly detail the key price-based DR programs.

#### 2.3.1.1 Time-Of-Use Pricing

TOU pricing is a price framework where electric power is priced differently based on time or seasonal variations [12]. Under the conventional TOU program, electricity is priced differently at peak periods and off-peak periods, depending on when electricity demand typically peaks. Some countries even have peak periods, off-peak periods and standard periods. Under a seasonal TOU pricing strategy, electric power is priced differently based on the seasons. Thus, seasons with higher demand for electricity, i.e. summer and winter, will have a different pricing scheme to seasons with lower demand, i.e. fall and spring. TOU programs are typically optional and have criteria for participating customers, with most consumers being industrial consumers.

#### 2.3.1.2 Real-Time Pricing

RTP frameworks provide a pricing strategy where electricity is more aligned to (near) real-time market conditions as opposed to TOU pricing frameworks that are predetermined based on prior time/seasonal prices [12]. It is often difficult to obtain practical real-time prices, so in practice prices are given to consumers from as close as a day ahead to an hour ahead of real time [10]. These prices are based on short-term forecasts of power system operational and market conditions.

#### 2.3.1.3 Critical Peak Pricing

In a bid to avert the hourly variation obtainable from RTP over a significant portion of operating hours, CPP is a pricing framework that attempts to model near "real"-time market conditions only at specific shorter time intervals/operating hours [12]. Thus, RTP is only deployed at those hours when the power system is extremely highly challenged and this may range from a few hours to a few days in a year, depending on the robustness and reliability of the power system. Consumers that sign up for the CPP program will receive notice from the utility on when the power system is expected to be significantly challenged (this may range from a few hours to a day) and the consumer is expected to take corresponding action by either curtailing load demand or paying for the more expensive electricity.

#### 2.3.1.4 Price Elasticity of Electrical Demand

Price elasticity is a metric used to measure the amount of change in consumer demand due to a price change. The conventional notion is that consumers' demand for goods or services reduces as the price increases. This magnitude of this change is referred to as the elasticity co-efficient. The mathematical formula for elasticity is given below:

$$E = \frac{\Delta d/d_0}{\Delta p/p_0}, \tag{2.1}$$

$$E = \frac{\Delta d p_0}{\Delta p d_0}, \tag{2.2}$$

where $\Delta d$, $\Delta p$, $d_0$, and $p_0$ are the change in demand, change in price, initial demand, and initial price, respectively [6]. Equations 2.1 and 2.2 are termed the self-elasticities, as they measure the change in demand for specific goods/services based on a price change for those goods/services. If, on the other hand, one wants to obtain the change in commodity of goods/services based on price change (increase/decrease) of another commodity, then this is referred to as cross-elasticities. The values for self-elasticities are usually −ve and +ve for cross-elasticities.

Deploying this concept to electricity markets, if a price increase in peak periods therefore causes a change in consumer demand in peak periods, then we measure this with self-elasticities and if this same price increase in peak periods leads to a change in demand in off-peak periods, we measure this with cross-elasticities. The mathematical formula for cross-elasticities is given below:

$$E(i,j) = \frac{\Delta d_i/d_{0i}}{\Delta p_j/p_{0j}} \tag{2.3}$$

$$E(i,j) = \frac{\Delta d_i p_{0j}}{\Delta p_j d_{0i}} \tag{2.4}$$

where $i$ and $j$ represent differing time frames such as off-peak and peak, $d_{0i}$ is the original consumer demand during $i$, $p_{0j}$ is the initial electricity price during $j$, $\Delta p_j$ is the price variation during $j$ and $\Delta d_i$ is the consumer demand variation during $i$. Assuming a daily scheduling interval with various peak and off-peak time periods, we use an elasticity matrix to represent the various changes in consumer demand to price changes (increase/decrease). The +ve off diagonal elements depict the cross-elasticities and the −ve diagonal elements depict the self-elasticities. Equation 2.5 gives the structure of an elasticity matrix for a given market over a 24-h time horizon.

$$E = \begin{bmatrix} E(1,1) & E(1,2) & \ldots & \ldots & E(1,24) \\ E(2,1) & E(2,2) & \ldots & \ldots & E(2,24) \\ \ldots & \ldots & E(i,j) & \ldots & \ldots \\ \ldots & \ldots & \ldots & \ldots & \ldots \\ E(24,1) & E(24,2) & \ldots & \ldots & E(24,24) \end{bmatrix} \tag{2.5}$$

#### 2.3.1.5 Determining Price Elasticity for Time-Of-Use Demand Response Programs

Using the concept of price elasticity matrices, it is possible to determine the load economic profile for price-based DR programs. An example is shown below on how to use price elasticity matrices to determine the load economic profile for the TOU DR program[6]:

We begin by defining $B(d_i)$ as the total consumer benefit in time interval $i$ that accrues from the utilization of $d_i$ kWh of energy and $p_i$ as the electricity price during interval hour $i$. The consumers' profit is denoted by:

$$S_i = B(d_i) - p_i d_i \tag{2.6}$$

For consumers' profit maximization, $\partial S_i / \partial d_i$ should be zero and we thus obtain:

$$\frac{\partial B(d_i)}{\partial d_i} = p_i \tag{2.7}$$

If we utilize the quadratic benefit function (any other benefit function can be used), then the total consumer benefit can be represented thus [6]:

$$B(d_i) = B(d_{0i}) + p_{0i}(d_i - d_{0i})\left[1 + \frac{d_i - d_{0i}}{2E(i,i).d_{0i}}\right] \tag{2.8}$$

where

$d_{0i}$ is the initial consumer demand during *i, i = 1,2,...,24;*
$p_{0i}$ is the initial electricity price during *i, i = 1,2,...,24;*
$E(i,i)$ is the self-elasticity;
$B(d_{0i})$ is the consumer benefit at $d_{0i}$;

$$\frac{\partial B(d_i)}{\partial d_i} = p_{0i}\left[1 + \frac{d_i - d_{0i}}{E(i,i).d_{0i}}\right]. \tag{2.9}$$

Equating (2.7) to (2.9), it is determined that:

$$d_i = d_{0i}\left[1 + \frac{E(i,i)[p_i - p_{0i}]}{p_{0i}}\right]. \tag{2.10}$$

Assuming multi-period elastic loads with consumers being able to reschedule, the total demand at interval *i* is a function of electricity price variations during *i = 1, 2... T*. Assuming further that the linearity assumption $\frac{\Delta d_i}{\Delta p_j}$ is constant for *i, j = 1, 2, 3,...0.24,* the relationship between electricity demand and price for a 24-h TOU program is:

$$d_i = d_{0i}\left[1 + \frac{\sum_{j=1}^{24} E(i,j)[p_j - p_{0j}]}{p_{0j}}\right]. \tag{2.11}$$

Combining Eq. 2.10 and 2.11 into one equation gives:

$$d_i = d_{0i}\left[1 + \frac{E(i,i)[p_i - p_{0i}]}{p_{0i}} + \frac{\sum_{j=1}^{24} E(i,j)[p_j - p_{0j}]}{p_{0j}}\right]. \tag{2.12}$$

#### 2.3.1.6 Appliance Scheduling Using Time-Of-Use Demand Response

In [6], a TOU DR framework is presented. There are three classes of DR loads: flexible loads, inflexible loads, and night-time loads. Inflexible loads are defined as those loads that affect the comfort of consumers and thus must always be switched on. Examples for residential consumers are space/water-heating technologies and perhaps cooking and refrigerating equipment. Some industrial loads are also inflexible, especially for critical industrial processes.

Flexible loads are those loads for which customers can easily adjust their demand in response to price variations. For residential consumers, examples include electric food-processing equipment, etc. Nighttime loads are defined as off-peak leads, which are often time intervals with the lowest tariffs. For residential consumers, this could be washing machines and dryers. Each load class has three different price elasticity matrices for a 24-h scheduling interval. The matrices are 24 × 24 square matrices and the difference between the three is based on the position of the non-zero elements. Equations 2.13–2.15 show the three matrices.

$$E_{\text{inf lexible}} = \begin{bmatrix} E(1,1) & E(1,2) & 0 & 0 & 0 & 0 & 0 \\ E(2,1) & E(2,2) & 0 & 0 & 0 & 0 & 0 \\ 0 & 0 & E(3,3) & E(3,j) & 0 & 0 & 0 \\ 0 & 0 & E(i,3) & E(i,j) & E(i,22) & 0 & 0 \\ 0 & 0 & 0 & E(22,j) & E(22,22) & 0 & 0 \\ 0 & 0 & 0 & 0 & E(23,22) & E(23,23) & E(23,24) \\ 0 & 0 & 0 & 0 & 0 & E(24,23) & E(24,24) \end{bmatrix}, \tag{2.13}$$

$$E_{\text{flexible}} = \begin{bmatrix} E(1,1) & 0 & 0 & 0 & 0 & 0 & 0 \\ E(2,1) & E(2,2) & 0 & 0 & 0 & 0 & 0 \\ E(3,1) & E(3,2) & E(3,3) & 0 & 0 & 0 & 0 \\ E(4,1) & E(4,2) & E(4,3) & E(4,j) & 0 & 0 & 0 \\ E(5,1) & E(5,2) & E(5,3) & E(5,j) & 0 & 0 & 0 \\ 0 & E(6,2) & E(6,3) & E(6,j) & 0 & 0 & 0 \\ 0 & 0 & E(7,3) & E(7,j) & 0 & 0 & 0 \\ 0 & 0 & 0 & E(8,j) & 0 & 0 & 0 \\ 0 & 0 & 0 & 0 & E(19,22) & E(19,23) & 0 \\ 0 & 0 & 0 & 0 & E(20,22) & E(20,23) & E(20,24) \\ 0 & 0 & 0 & 0 & E(21,22) & E(21,23) & E(21,24) \\ 0 & 0 & 0 & 0 & E(22,22) & E(22,23) & E(22,24) \\ 0 & 0 & 0 & 0 & 0 & E(23,23) & E(23,24) \\ 0 & 0 & 0 & 0 & 0 & 0 & E(24,24) \end{bmatrix}, \tag{2.14}$$

$$E_{\text{night-time}} = \begin{bmatrix} E(1,1) & E(1,2) & \ldots & E(1,j) & \ldots & E(1,23) & E(1,24) \\ E(1,2). & E(2,2) & \ldots & E(2,j) & \ldots & E(2,23) & E(2,24) \\ 0 & 0 & E(3,3) & 0 & 0 & 0 & 0 \\ 0 & 0 & 0 & E(i,j) & 0 & 0 & 0 \\ 0 & 0 & 0 & 0 & E(22,22) & 0 & 0 \\ E(1,23) & E(2,23) & \ldots & E(i,23) & \ldots & E(23,23) & E(23,24) \\ E(1,24) & E(2,24) & \ldots & E(i,24) & \ldots & E(23,24) & E(24,24) \end{bmatrix}. \tag{2.15}$$

The above information can be used to obtain an optimal minimal cost schedule for inflexible, flexible and night-time loads under a TOU tariff plan [8]. This is referred to as appliance scheduling.

## 2.4 Incentive-Based Demand Response

When using IB-DR programs, electricity consumers shut down or curtail their electricity load and in return receive incentives for doing so. An overview of major IB-DR programs is detailed below:

### *2.4.1 Direct Load Control*

DLC programs allow the utility supplier to leverage on advanced monitoring and control technologies to switch off or curtail the operations of specific customer equipment in a bid to provide relief to the power grid [10]. The customers are typically residential or smaller industrial or commercial organizations. In return, the consumers are given either a monetary rebate or a reduced electricity tariff as their compensation for participation. Typically, the consumers do not have control over their participation in DLC programs. There are, however, some DLC programs where consumers retain control over their equipment and can choose to decline to participate when given the signal to curtail energy consumption. In such a case, consumers are penalized depending on the permissible number of times they may choose to decline participation. DLC programs depend on excellent communication infrastructure for proper functioning. DLC is typically applied to energy-guzzling appliances such as water- and space-heating equipment.

### *2.4.2 Interruptible Programs*

Interruptible programs are geared towards large-scale energy consumers' (loads > = 1 MW) in order to incentivize them to discontinue their operations for a pre-determined period of time [12]. These large-scale energy consumers can be activated quickly in the event the power system becomes unexpectedly stressed or strained and either have their own back-up power-generation schemes or can shift their operational schedule to when the system is less stressed [12].

The incentive in these cases is rates, which are typically lower than the conventional tariffs, and these programs have strict terms, with severe penalties for non-compliance. The programs, for example, will explicitly spell out the terms of the interruption, quantity of power to be shut down by participating consumers, reduced rates, which are the incentives for compliance, and the penalties for non-compliance.

### 2.4.3 Curtailable Load Program

The curtailable load program is essentially a variant of the interruptible program through which energy consumers are again incentivized to provide power system relief. However, unlike the interruptible program, these companies are not requested to discontinue their operations completely, but to reduce or curtail their energy consumption [12]. Another key difference from the interruptible program is that the consumers do not necessarily have to be large-scale consumers; the program often involves a mix of large-scale and medium-to-large-scale consumers. Yet another difference is that typically the penalties for curtailable load programs and the window frame for compliance are lower than those of the interruptible program. Furthermore, the nature of the program allows consumers to choose the devices/loads to be reduced in order to achieve the previously agreed target for load reduction, as this is not enforced by the utility, unlike the DLC program.

### 2.4.4 Demand-Bidding Programs

In demand-bidding programs, consumers are incentivized to curtail their electricity load demand [12]. Two formats are used for this program. The first is the customer-bidding program and the second is termed sponsor bidding. In the customer-bidding program, consumers are the active bidders and submit a day-ahead bid to the utility, indicating the amount of load they are willing to curtail and at what price. The utility then considers the bids received in addition to the forecast power system condition and decides on the bids to be accepted [12].

In the sponsored bidding program, the utility is the active participant and announces a predetermined incentive (price) it is willing to pay for demand curtailment. Electricity consumers then consider the amount and in return indicate the quantum of load they are willing to curtail, based on the utility's announced price. Under both programs, the consumer is thereafter paid upon fulfillment of its own side of the deal [12].

### 2.4.5 Game Theory Demand Response Programs

A critical issue determined in IB-DR programs is the amount/quantity of the incentive offered by the utility. If the incentive is too little, consumers will not be interested in participating [13–15]. On the other hand, the utility cannot afford to give excessively high incentives because of budget and affordability issues. The utility therefore has to offer incentives that will truly attract consumer participation, which means that the incentive should ideally be greater than the cost of power interruption. To combat this challenge, game theory has been deployed to design optimal DR programs.

In references [13, 14], the authors delineate three key criteria that incentive-based demand management programs should consider:

- IB-DR programs should distinguish between different customers. Different customers imply dissimilar loads and dissimilar load outage costs.
- In order to offer attractive consumer incentives, the utility should be able to determine in monetary terms ballpark ranges of what these consumer load outage costs amount to.
- Incentive-based programs also have to factor in the locations of the consumers in calculating incentives, as different consumers have different impacts on the power grid.

An optimal approach to designing IB-DR programs while incorporating all three design criteria is detailed below.

#### 2.4.5.1 Game Theory Based Demand Response Formulations

We begin by defining key parameters:

- $\theta$ differentiates different kinds of customers and is normalized between 0 and 1 ($0 \leq \theta \leq 1$) with $\theta = 1$ depicting the consumer most inclined to shed power and $\theta = 0$ depicting the consumer least inclined to shed electric power.
- $x$ is the amount of power either curtailed or shut down by an electricity consumer.
- $c(x, \theta)$ is customer $\theta$ incurred cost due to load reduction/curtailment of $x$ MW
- $\lambda$ factors in the locational impact different consumers have on the grid and depicts in monetary terms the cost of failure to supply power to particular consumers based on their location on the grid. In the event of the grid being strained, it might be better for the utility not to supply certain locations/buses and the utility can determine the value typically from an optimal power flow (OPF) calculation. This is also referred to as the "value of power interruptibility" ($\lambda$) [13, 14].

We can give the benefit function of the consumer as:

$$V_1(\theta, x, y) = y - c(\theta, x) \tag{2.16}$$

where y is the incentive received by the consumer for shedding load. In order to ensure attractiveness to the consumer, $V_1 \geq 0$.

In the same vein, the utility's benefit function is given as:

$$V_2(\theta, \lambda) = \lambda x - y \tag{2.17}$$

The utility's aim will therefore be to maximize its benefit:

$$\max_{x,y}[\lambda x - y]. \tag{2.18}$$

The assumed form of the consumers cost function is given below:

$$c(\theta, x) = K_1 x^2 + K_2 x - K_2 x\theta \tag{2.19}$$

where $K_1$ and $K_2$ are the cost function coefficients. Equation 2.19 is a quadratic cost function. The cubic and exponential cost functions are other kinds of cost functions also used [13]. However, cost functions must satisfy the following criteria [13, 14]:

- An increase of $\theta$ decreases marginal cost and increases marginal benefit.
- The marginal cost should be non-negative.
- The cost function should be convex (increasing marginal cost).
- If a customer does not reduce use or curtail any power, the cost should be zero ( $c(\theta, 0) = 0$).
- $c(\theta, x) = K_1 x^2 + K_2 x - K_2 x\theta$ (quadratic cost function).
- The $K_2 x\theta$ term sorts customers by way of $\theta$.
- $\frac{\partial c}{\partial x} = 2K_1 x + K_2 - K_2\theta$.

Assuming there are J customers and one utility, $y_j$ is the monetary value of payments to customer $j$ with the utility benefit for customer $j$

$$u_j = y_j - (K_1 x_j^2 + K_2 x_j - K_2 x_j \theta_j), \text{for} \quad j = 1,\ldots,J. \tag{2.20}$$

Therefore, the total utility benefit for $J$ customers is:

$$u_0 = \sum_{j=1}^{J} \lambda_j x_j - y_j \tag{2.21}$$

The utility will want to maximize this benefit.

$$\max_{x,y} \sum_{j=1}^{J} [\lambda_j x_j - y_j] \tag{2.22}$$

The utility will also want to enforce two requirements, termed the "*individual rationality constraint*" and the "*incentive compatibility constraint*", both of which are represented below:

$$y_j - (K_1 x_j^2 + K_2 x_j - K_2 x_j \theta_j) \geq 0, \text{ for } \quad j = 1,\ldots,J. \tag{2.23}$$

$$y_j - (K_1x_j^2 + K_2x_j - K_2x_j\theta_j) \geq y_{j-1} - (K_1x_{j-1}^2 + K_2x_{j-1} - K_2x_{j-1}\theta_{j-1}), \text{for } j = 1, \ldots, J. \tag{2.24}$$

The "*individual rationality constraint*" seeks to enforce the requirement that each consumer has a +ve benefit while the "*incentive compatibility constraint*" enforces compensation corresponding to the amount of power shed.

## 2.5 Summary

In this chapter, DSM programs encompassing energy efficiency and DR programs were detailed. DR programs are subdivided into incentive-based and price-based programs, which were considered in detail. Examples of energy efficiency programs (a retrofit program), price-based DR (TOU appliance scheduling) and IB-DR programs (game theory-based incentive-based DR) were discussed in detail and their mathematical formulations were presented.

## References

1. Wirl F (1997) The economics of conservation programs. Kluwer Academic Publishers, Dordrecht, The Netherlands
2. Lovins A (1985) Saving gigabucks with negawatts. Public Utilities Fortnightly 115(6):19–26
3. North American Electric Reliability Corporation (2011) Demand response availability report, March 2013
4. Nwulu N, Sun Y "Optimal incentivized demand response and building retrofits", 3rd International conference on advances in computing, communication and engineering, ICACCE 2016, Durban, South Africa 28, 29 November 2016
5. Department of Energy (Feb 2006) U. S. benefits of demand response in electricity markets and recommendations for achieving them
6. Nwulu NI, Xia X (2015) A combined dynamic economic emission dispatch and time of use demand response mathematical modelling framework. J Renew Sustain Energ 7(4)
7. Nwulu NI, Fahrioglu M (2011) Power system demand management contract design: a comparison between game theory and artificial neural networks. Int Rev Model Simul 4 (1):104–112
8. Nwulu NI, Fahrioglu M (8–11 May 2011) "A neural network model for optimal demand management contract design", Proceedings of the 10th international conference on environmental and electrical engineering (EEEIC2011). Rome, Italy
9. Nwulu NI, Xia X (2015) Multi-objective dynamic economic emission dispatch of electric power generation integrated with game theory based demand response programs. Energ Convers Manage 89(1):963–974
10. National Action Plan for Energy Efficiency (2010) Coordination of energy efficiency and demand response. Prepared by Charles Goldman (Lawrence Berkeley National Laboratory), Michael Reid (E Source), Roger Levy, and Alison Silverstein
11. Fahrioglu M, Nwulu NI (Nov 2012) Investigating a ranking of loads in avoiding potential power system outages. J Electr Rev (Przeglad Elektrotechniczny), Warsaw, Poland 88(11a): 239–242

12. Rocky Mountain Institutie (30 April 2006) Demand response: an introduction, overview of programs, technologies and lessons learned. Southwest Energy Efficiency Project
13. Fahrioglu M, Alvarado FL (May 2001) Using utility information to calibrate customer demand management behavior models. IEEE Trans Power Syst 16(2)
14. Fahrioglu M, Alvarado FL (Nov 2000) Designing incentive compatible contracts for effective demand management. IEEE Trans Power Syst 15(4)
15. Damisa U, Nwulu NI, Sun Y (2018) Microgrid energy and reserve management incorporating prosumer behind the-meter Resources. IET Renew Power Gener 12(8):910–919

# Chapter 3
# Mathematical Optimization Modeling and Solution Approaches

## 3.1 Introduction

Mathematical optimization is a branch of applied mathematics that is concerned with determining the best option among a collection of options with respect to a clearly defined objective or yardstick. The word optimization is derived from the word optimal, which can loosely be translated to mean best, peak, prime. Thus, mathematical optimization is mathematically determining the best option or alternative from a number of choices.

Mathematical optimization modeling has found application in diverse fields, subject matters, and industries. It is widely used in engineering, supply chain modeling, finance, and medicine; in fact, in just about any field of human endeavor.

A mathematical optimization model consists of three critical elements. The first is a well-defined objective, which the modeler wants either to maximize or minimize. In the terminology of mathematical optimization, this is referred to as the objective function. The second is variables, which are elements that affect one's objective and that one is able to vary. The last critical element are constraints, which are mathematical equations that impose relevant limits on one's variables. All three, objectives, variables, and constraints, can be abstracted from any problem of interest in any domain to find the "optimal" or "best" option among a set of competing alternatives. This abstraction process is referred to as mathematical modeling and it has become a scientific endeavor owing to the need to deploy mathematical optimization in very complicated and complex domains. The resulting set of equations (consisting of objectives, variables, and constraints) is known as a mathematical model.

A mathematical model is typically written in the general form:

$$\text{Minimize or Maximize}\, F(x) \tag{3.1}$$



N. Nwulu and S. L. Gbadamosi, *Optimal Operation and Control of Power Systems Using an Algebraic Modelling Language*, Green Energy and Technology,
https://doi.org/10.1007/978-3-030-00395-1_3

$$\begin{aligned} &\text{Subject to} \\ &g(x) < = b \end{aligned} \quad (3.2)$$

$$h(x) = c \quad (3.3)$$

Equation (3.1) is the objective function, Eq. 3.2 is the inequality constraint and Eq. 3.3 is the equality constraint.

Mathematical optimization uses computing machinery to solve the resulting model and requires a solver, which simply put is a mechanism for reading the mathematical model that provides a practical solution.

There are various classifications of mathematical optimization models. A common classification is based on the presence of constraints. Thus, optimization models with constraints are referred to as constrained optimization models, while models without constraints are called unconstrained models. It is worth noting that most real-world problems are constrained problems. Another classification is based on the type of variable. Variables can either be discrete or continuous. This therefore translates into discrete or continuous mathematical optimization problems. Yet another classification is based on the linearity of the objective function or constraints. We therefore have linear optimization models (commonly referred to as linear programming) or nonlinear optimization models (nonlinear programming). Other common classifications of mathematical models include static or dynamic models, which distinguish optimization models that include time variance. There are also stochastic or deterministic mathematic models, which distinguish models that contain elements of randomness and uncertainty. We also have single-objective or multi-objective optimization models, which distinguish models, based on the number of objectives that the modeler wants to maximize or minimize.

There are many mathematical optimization solution techniques and this has become an oft-researched scientific area. Figure 3.1 shows a high-level overview of different mathematical optimization solution techniques. In this work, we categorize these mathematical optimization solution approaches into classical mathematical optimization solution approaches/techniques and meta-heuristic solution approaches/techniques.

## 3.2 Classical Mathematical Optimization Solution Approaches

Classical mathematical optimization solution approaches can be divided further into two classes, based on their solution methodology, namely, gradient descent methods and direct search methods [2]. Both approaches are explained further in the following subsections.

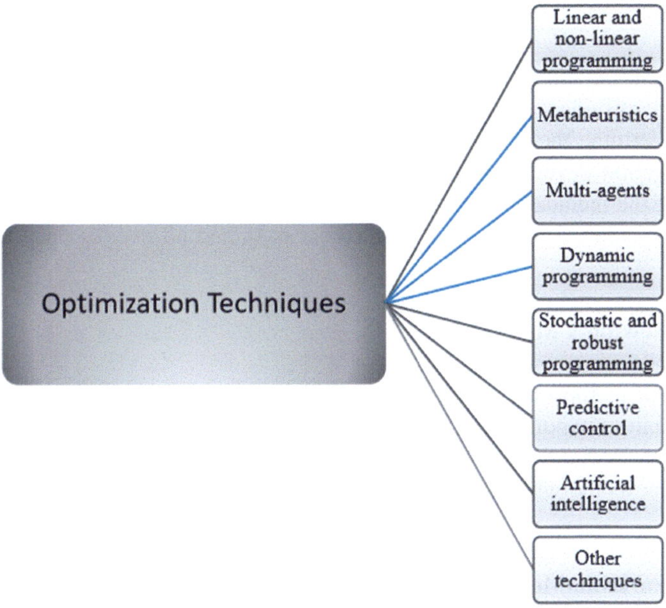


**Fig. 3.1** An overview of mathematical optimization techniques [1]

## *3.2.1 Gradient Descent-Based Methods*

Gradient descent-based methods are generally regarded as more proficient than other methods. This approach leverages on calculus and the determination of the derivatives of the objective function and constraints to obtain solutions [2]. An overview of some gradient descent solution methods is given below:

### 3.2.1.1 Steepest Descent Method

The steepest descent method starts with a guess for the minimum point and determines the direction of the steepest point from the initial minimum point [2]. If the direction obtained is not equal to zero (i.e. convergence has not yet been reached), then a line search for the minimum point is done along the negative to derivative direction [2]. The minimum point obtained then becomes the current point and the process is repeated till a zero value is obtained (i.e. convergence is obtained).

#### 3.2.1.2 Conjugate Gradient Method

The conjugate gradient method is a variant of the steepest descent method and works by modifying the search direction [2]. In this method, recursive equations are used to determine the search direction. Because of this, after a few iterations, the search direction develops linear dependence. The degree of this is determined by calculating the included angle of two successive search directions. In the event that this value is approximately zero, the process is started all over again [2].

#### 3.2.1.3 Newton's Method

The Newton method uses second-order derivatives to determine search directions for the minimum point [2]. Second-order Taylor's series expansion is used to determine the value of the function around the current point and the global minima are denoted, as the where the Hessian matrix is positively semi-definite [2].

#### 3.2.1.4 Variable-Metric Method

The variable-metric method uses first-order derivatives to approximate the inverse of the Hessian matrix, thereby bypassing computation of the Hessian matrix and its inverse.

### 3.2.2 *Direct Search Methods*

Direct search methods do not use the function's partial derivatives in determining the search direction. The search for the minimum is only conducted using the value of the function at different points, thus this method is usually suitable for less complicated problems with relatively few variables [2]. An overview of some gradient descent solution methods is given below:

#### 3.2.2.1 Hooke-Jeeves Method

The Hooke-Jeeves method is made up of two moves: the exploratory and the pattern move [2]. The exploratory move is first done in the region around the current point to determine the best point, after which the pattern search is conducted using the two previously identified points.

#### 3.2.2.2 Powell's Conjugate Direction Method

Powell's method uses results from prior solutions to develop new search directions. It uses the history of previous searches to develop a number of linearly independent directions and conducts searches in these directions [2]. It is an efficient search method.

## 3.3 Meta-Heuristic Methods

Meta-heuristic methods are indicated in Fig. 3.2 before being discussed in more detail.

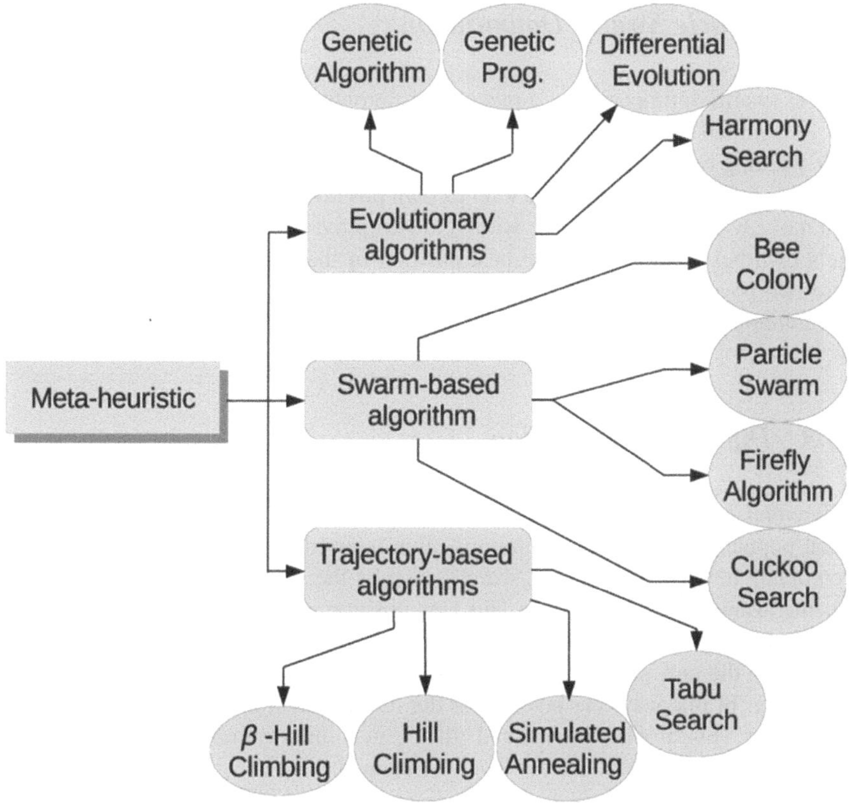


**Fig. 3.2** An overview of meta-heuristic optimization solution methods [3]

### 3.3.1 Genetic Algorithm

A genetic algorithm is a search algorithm that mimics the theory of genetics and natural selection from human nature as conceptualized by Darwin's theory of evolution. The algorithm works by first randomly initializing an initial population of candidate solutions (with alterable chromosomes) [4]. The fitness level of this initial generation is evaluated. This population is thereafter refined first by the reproduction operator, which selects the best solutions from the initial random population and forms the mating pool. Once again, the fitness of this generation is evaluated. The cross-over operator is then used on solutions in the mating pool, after which the mutation operator is further used to obtain another generation whose fitness is yet again evaluated. The process continues until either the prerequisite fitness level is obtained or the maximum number of generations has been obtained.

### 3.3.2 Particle Swarm Optimization

Particle swarm optimization is a meta-heuristic optimization solution methodology that is regarded as a heuristic optimization procedure based on the movement of birds flocking or fish schooling [4]. Each individual (in either the bird flock or fish school) is referred to as a particle with its own position and velocity. These particles then move within a search space with guidance provided by the knowledge of each one's own best position and the best position of the entire group. Knowledge of these two positions, known as local best and global best, respectively, enables the best solutions to be obtained.

### 3.3.3 Differential Evolution

Differential evolution is a meta-heuristic solution approach that begins with an initial population of potential solutions often referred to as agents [4]. The agents then obtain new positions in the search space by the weighted difference vector of existing agents in the population. This new position is evaluated and if its score (in relation to a fitness function) is an improvement on its previous position, it is included in the new population, otherwise the new agent is excluded from the new population. In this way, the position of the new population is an improvement on the former. This process is repeated till the maximum number of populations is obtained or a manageable fitness level is obtained.

### 3.3.4 Group Search Optimization

Group search optimization is a meta-heuristic solution methodology that uses concepts related to the way animals search for food [4]. The initial population is selected and each member of this population can be a producer, scrounger or a ranger and mimics the way a herd of animals produces, scrounges or disperses (ranges) in a bid for sustenance. Each member of the population is thereafter evaluated based on its activities and its resultant position in relation to the fitness function. The process continues until a predefined termination criterion is obtained (Fig. 3.3).

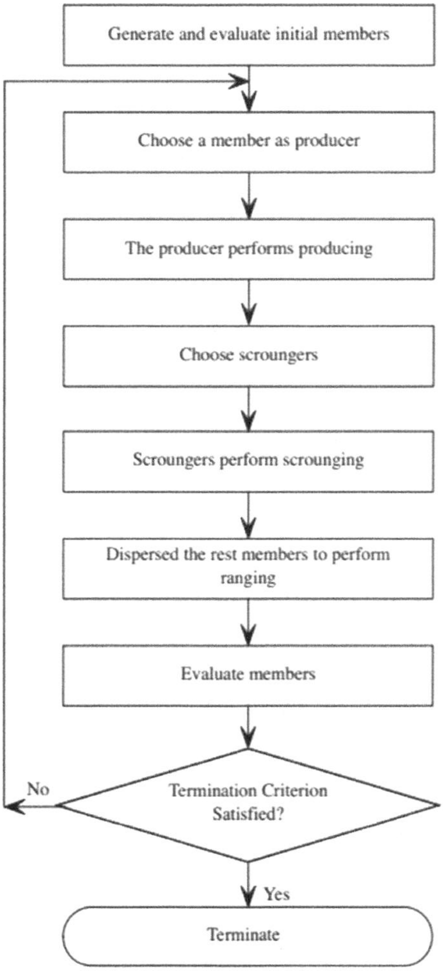


**Fig. 3.3** Flow chart for group search optimization [5]

### 3.3.5 *Cuckoo Search*

The cuckoo search method is a meta-heuristic solution methodology that is motivated by the way cuckoo birds brood their offspring [4]. Cuckoo birds lay one egg at a time and deposit the egg in a nest (the choice of this nest is random). An egg in a nest depicts a solution. After a while nests start filling up, with each new egg depicting a new solution. The nests with the largest collection of good quality eggs are carried over into the next generation and nests without good eggs are discarded. This process is continued until the predetermined termination criteria are reached.

### 3.3.6 *Grey Wolf Optimization*

This solution methodology is an algorithm that is based on the societal norms and hunting behavior of a pack of grey wolves [4]. In terms of societal hierarchy, grey wolves have alpha, beta, delta, and omega wolves in order of hierarchy. This societal hierarchy is adhered to when they hunt prey. The hunting activities of the pack consist of identifying a prey, encircling, and attacking the prey. The wolves in the pack are assumed to be potential solutions, with the hierarchy also determining the quality of solutions. Thus, the alpha wolves lead the pack to search for the prey, which, after it has been identified, is pursued and encircled. At that point the search process ends and the attack begins. The positions of all the solutions (wolves) are updated throughout the hunting procedure and their fitness is evaluated in relation to the fitness (location) of the prey. This process continues till the predetermined termination criteria have been reached.

### 3.3.7 *Harmony Search Algorithm*

The harmony search algorithm is a meta-heuristic solution algorithm based on the way musicians improve the harmony of their music [4]. A musician improves the harmony of his music by playing from memory (a harmony previously memorized), adjusting the pitch (modifying an existing memorized harmony) or making a randomized attempt. If the harmony is a candidate solution to a problem, then an optimal solution (perfect harmony) is obtained by initially obtaining previously memorized harmonies (solutions) and then modifying/adjusting them and finally comparing them with the randomized harmonics. Depending on which harmonics are better, the iteration continues. The aim of randomized harmonies is to force a search of space further afield. This process continues till a predetermined termination criterion is obtained.

### 3.3.8 *Bee Colony Optimization*

Bee colony optimization is a meta-heuristic optimization methodology that mimics the way bees forage for food [4]. The bees are denoted as solutions; there are two types of bees: scouting and employed bees, with scouting bees searching for new food sources while employed bees obtain food from previously established food sources. Both classes of bees update themselves with information about their positions and abandon locations where food is not found. This process continues until either food is found or a predetermined termination criterion is achieved.

## 3.4 The Mathematical Modeling Process

The optimization modeling process is a cyclical process involving a number of steps. The primary and most critical step is understanding the nature of the real-world problem and the various factors (controllable or not) that influence the problem. What needs to be minimized/maximized and what are the factors that influence this quantity to be minimized/maximized? What factors can be ignored/ eliminated or approximated? The second step is the mathematical representation of the previously understood real-world problem. This involves confirming that the mathematical model obtained is a correct representation of the problem and any assumptions are valid/correct without distracting from the physical problem at hand [6, 7]. The third step involves obtaining data and processing the data that are required. This is followed by representing the mathematical model in syntax that the solver can understand, running the solver, and obtaining results. It is envisaged that the results should ideally be optimal. These results are subsequently analyzed in order to glean actionable intelligence from the results. The modeling process just described is an ideal process where everything goes according to plan, which hardly happens in modeling practice. In practice, a number of bottlenecks often occur, which will lead to a revision of some of the modeling assumptions, a review of data used, and/or refinement of algorithms or solvers deployed. This makes modeling and optimization of mathematical modeling a very cumbersome process.

In practice, these are the steps to be used when building and solving a mathematical optimization model:

i. Problem conceptualization
   The first step is problem conceptualization, which involves determining the problem, factors that influence the problem, and the mathematical abstraction of the problem.

ii. Mathematical model design
This step involves the conversion of the mathematical problem that has been determined into a form that can be solved by the computer. This is dependent on the platform used, solver, etc.
iii. Results and analysis
After running the software or solver and obtaining results, the results are analyzed to determine their accuracy and reasonableness.
iv. Review
After analyzing the results, especially in cases where shortfalls occur, the model is reviewed and appropriate corrective action is taken. If the results are accurate and reasonable, actionable intelligence is gleaned from the work.

## 3.5 Algebraic Modeling Languages

AMLs are advanced software platforms that enable the solution of large-scale complex optimization problems. AMLs are a bridge between the high-level algebraic mathematical representation of optimization problems and the low-level algorithms that enable their solution. AMLs use a syntax that is analogous to the mathematical symbolization of optimization problems, thereby substantially eliminating human effort and time in the optimization modeling process [8]. This introduces efficiency into the modeling process and more importantly, allows separation of the mathematical model, the data, and the solver. This separation allows for a number of different experiments and is one of the key concepts that have led to increased use of AMLs. Thus, the mathematical model and solver can be kept constant and the data can be scaled either up or down to represent larger or more realistic scenarios. In addition, the mathematical model and the date can be left constant and different solvers can experimented with in a bid to obtain the true optimum. The final choice of solvers used can be based on accuracy (true optimality), processing speed, and other relevant performance metrics.

### *3.5.1 Software*

AMLs are high-level modeling platforms that can be used on computers. They contain integrated development environments and data exchange capabilities and are connected to a large number of algorithms that can be used to solve optimization problems. These algorithms are typically referred to as solvers and can either be open-source or commercial. AML software can be installed on most types of computers and operating systems and are deployed across a wide range of industries and businesses.

There are many AMLs. However, five are adjudged to be the most important ones [9]:

- Advanced Interactive Multidimensional Modeling System (AIMMS), founded by Johannes Bisschop
- General Algebraic Modeling System, founded by Johannes Bisschop and Alexander Meeraus
- Mathematical Programming Language, founded by Bjarni Kristjansson
- A Mathematical Programming Language, founded by Robert Fourer, David Gay, and Brian Kernighan
- Linear, Interactive, and Discrete Optimizer, founded by Kevin Cunningham and Linus Schrage.

Most AMLs are quite similar in design philosophy; however, differences exist in their syntax, declarative and procedural elements, solver links, etc. AMLs can also be integrated with conventional mathematical and analytical platforms such as MATLAB and Microsoft Excel [8].

AMLs can also be deployed in software as a service framework. An example is the Network-enabled Optimization System server, which gives mathematical modelers access to a large number of solvers [8]. The server allows for a solution of 12 different kinds of optimization problems. This unique service allows for the sharing of optimization resources across a wide range of users.

As stated before, algorithms that can be used to solve optimization problems are referred to as solvers. Solvers typically differ based on the type of optimization problems they can solve and whether they are free, commercial, or open-source. It is also possible to merge solvers, depending on the nature and complexity of the problem. Examples of prominent solvers include Sparse Nonlinear Optimizer, Computational Infrastructure for Operations Research, and IBM ILOG CPLEX Optimization Studio (CPLEX), among many others.

## 3.6 Advanced Interactive Multidimensional Modeling System

The AIMMS is an AML that has evolved into a prescriptive analytics platform that enables the creation of business-ready optimization applications. [10], [11]. It has also evolved to the AIMMS supply chain navigator, which provides applications for supply chain purposes. The software can solve a wide variety of optimization problems and includes an equally wide range of solvers [12]. A very attractive feature for academic institutions is the free academic license for academic institutions, which enables academics and students to access the software for research purposes.

### 3.6.1 Software Installation

AIMMS can be installed on various operating systems (Windows or Linux) and has many releases [10]. Except for academics and students, as stated, a license is required to use AIMMS. The process is intuitive and easy to follow.

AIMMS has a number of helpful tutorials for varying levels of expertise.

### 3.6.2 AIMMS Tutorials

The AIMMS tutorial for beginners, as the name implies, is for beginners and introduces AIMMS [11]. It shows how to build relatively fast yet powerful models and includes useful data-handling functionality. After this tutorial, there are the AIMMS tutorials for professionals, which cover advanced topics for experienced modelers, including rolling horizons, etc.

Practicalissuesto be addressed when using AIMMS:

i. Defining the sets and indices;
ii. Determining the parameters;
iii. Determining the decision variables;
iv. Defining the objective function
v. Defining constraints;
vi. Building the model on the chosen AML;
vii. Loading the data;
viii. Choosing the solver to use (AIMMS does this automatically but the user can still choose solvers);
ix. Obtaining results;
x. Analyzing results;
xi. Reviewing previously performed steps if results are not as expected.

In the next section, we briefly show how to use AIMMS for building retrofit energy efficiency programs, mentioned in 2.2.1, appliance scheduling using

TOU DR, mentioned in 2.3.1.5, and the game theory demand response programs, mentioned in 2.4.

## 3.7 Building Retrofit Energy Efficiency Programs Using AIMMS

The problem was introduced in Sect. 2.2.1. A complete description is provided in [13].

**Set Definition**:

$j$ represents customers and $i$ represents technologies to be retrofitted.

**Variable Definition**:

$n_{ji}$ is the number of units of technology $i$ to be replaced by the customer $j$.

**Parameter Definition**:

| | |
|---|---|
| $R_i$ | is the rebate paid by the utility for each technology $i$ in South African rands (ZAR). |
| $n_{ji}$min and $n_{ji}$max | are the minimum and maximum number of units of technology $i$ to be replaced by the customer $j$. |
| $C_i$ | is the average customer cost for replacing technology $i$ (ZAR). |
| $\Delta P_i$ | is the difference in power rating between the old technology and replacement technology (W). |
| $\lambda_j$ | is the electricity tariff for the customer $j$ in ZAR/kWh. |
| $h_{ji}$ | is the average number of hours that a customer $j$ uses technology $i$ in a day. |
| $MR$ | is the maximum rebate Eskom is paying for any project capped at ZAR 750,000, i.e., MR = 750,000. |
| $CB_j$ | is the maximum amount of money customer $j$ is willing to spend on replacing inefficient technologies. |
| $J$ | is the total number of customers. |
| $I$ | is the maximum number of technologies the customer selects from the standard product program and is willing to change. |

**Objective Function and Constraints**

$$\max f_2 = \sum_{j=1}^{J}\sum_{i=1}^{I}\left[n_{\mathrm{ji}}(R_i - C_i) + 365 n_{\mathrm{ji}} h_{\mathrm{ji}} \Delta P_i \lambda_j\right] \tag{3.4}$$

subject to:

$$\sum_{i=1}^{I} n_{\mathrm{ji}} R_i \leq \mathrm{MR}, \tag{3.5}$$

$$\sum_{i=1}^{I} n_{\mathrm{ji}} C_i \leq CB_j, \tag{3.6}$$

$$1\mathrm{kW} \leq \sum_{i=1}^{I} n_{\mathrm{ji}} \Delta P_i \leq 250\mathrm{kW}, \tag{3.7}$$

$$\sum_{i=1}^{I} 365 n_{\mathrm{ji}} h_{\mathrm{ji}} \Delta P_i \geq 2\mathrm{MWh}. \tag{3.8}$$

for all $j = 1,2,.. .. .., J$.

The following is a brief explanation of the constraints [9]:

- The first constraint (3.5) ensures that for each customer, the rebate paid is limited to ZAR750 000.
- The second constraint (3.6) limits the total customer investment to the individual customer's budgeted amount.
- The third constraint (3.7) ensures that the demand reduction for each customer due to the technology replacement is between 1 kW and 250 kW.
- The fourth constraint (3.8) compels the energy saved per annum to be greater than 2 MWh.

```
Model Main_Example3.7 {
    Set customers {
        Index: j;
    }
    Set technologies {
        Index: i;
    }
    Parameter Rebate {
        IndexDomain: (i);
    }
    Parameter customercost {
        IndexDomain: (i);
    }
    Parameter Poweratingdiff {
        IndexDomain: (i);
    }
    Parameter Tarriff {
        IndexDomain: (j);
    }
    Parameter hourofuse {
        IndexDomain: (j,i);
    }
    Parameter MaxRebate;
    Parameter Max_Replacementcost {
        IndexDomain: (j);
    }
    Parameter Min_technologyunit {
        IndexDomain: (j,i);
    }
    Parameter Max_technologyunit {
        IndexDomain: (j,i);
    }
    Variable no_technologyunit {
        IndexDomain: (j,i);
        Range: nonnegative;
    }
    Variable Obj_Function {
        Range: free;
```

```
        Definition: sum((j,i),(no_technologyunit(j,i)*(Rebate(i)-
customercost(i))))+sum((j,i),(365*(no_technologyunit(j,i)*(hourofuse(j,i)*(Poweratingd
iff(i))*(Tarriff(j))))));

        Comment: "sum((j,i),((no_technologyunit(j,i)*(Rebate(i)-
customercost(i)))+(365(no_technologyunit(j,i)*(hourofuse(j,i)*(Poweratingdiff(i))*(Tar
riff(j)))))))";

    }

    Constraint constr1 {

        IndexDomain: (j);

        Definition: sum(i,(no_technologyunit(j,i))*(Rebate(i)))<=MaxRebate;

    }

    Constraint constr2 {

        IndexDomain: (j);

        Definition:
sum(i,(no_technologyunit(j,i))*(Rebate(i)))<=Max_Replacementcost(j);

    }

    Constraint constr3 {

        IndexDomain: (j);

        Definition: 1<=sum(i,(no_technologyunit(j,i))*(Poweratingdiff(i)))<=250;

    }

    Constraint constr4 {

        IndexDomain: (j);

        Definition:
sum(i,365*(no_technologyunit(j,i))*(hourofuse(j,i))*(Poweratingdiff(i)))<=2000;

    }

    MathematicalProgram Net {

        Objective: Obj_Function;

        Direction: minimize;

        Constraints: AllConstraints;

        Variables: AllVariables;

        Type: Automatic;

    }

    Procedure MainInitialization {

        Comment: "Add initialization statements here that do NOT require any library
being initialized already.";

    }

    Procedure PostMainInitialization {

        Comment: {

            "Add initialization statements here that require that the libraries are
already initialized properly,

            or add statements that require the Data Management module to be
initialized."

        }
```

```
}
Procedure MainExecution {
    Body: {
        !solve NetCost;
    }
}
```

### 3.7.1 *Appliance Scheduling Using Time-of-Use Demand Response*

The problem was introduced in Sect. 2.3.1.5. A complete description is provided in [13].

**Set Definition**:

$t$ represents half-hourly time slots and $i$ represents the loads of the customer.

**Variable Definition**:

$u_{\mathrm{it}}$ shows if the load $i$ is switched ON or OFF in time $t$ (status of load $i$ in time slot. $t$)

**Parameter Definition**

$B_i$ energy consumption of load $i$ in each time slot(kWh)
$k$ TOU tariff at each time slot
$E_i$ end time slot for load $i$
$S_i$ start time slot for load $i$
$Z_i$ total number of time slots required for load $i$ to be switched ON
A is the total number of customers.

**Objective Function and Constraints**

$$\min \sum_{t=1}^{24} \sum_{i=1}^{I} B_i U_{\mathrm{it}} p_i, \tag{3.9}$$

subject to:

$$\sum_{t}^{24} V_{\mathrm{it}} = Z_i, \tag{3.10}$$

$$\sum_{t=1}^{24} B_i U_{\mathrm{it}} \leq E_i \tag{3.11}$$

where [10]

$B_i$ is the energy consumption of load $i$ in each time slot (kWh), respectively;
$U_{\mathrm{it}}$ is the decision variables either 1 or 0 and states if the load$i$is switched on or off in time slot $i$;
$p_i$ is utility defined price/tariff for each time slot in South African Rands ZAR/kWh;
$Z_i$ is the total number of time slots required for load$i$ to complete its task;
$S_i$ is the start time slot for load $i$;
$E_i$ is the end time slot for load $i$;

The following is a brief explanation of the constraints:

- Constraint (3.11) ensures that there are sufficient time slots for a load, device, or machine to complete its tasks. Also this constraint handles the flexibility of the appliance/load.
- The final constraint (3.12) ensures that the customer's new energy level does not exceed the previous allowable energy level of the customer.

```
Model Main_Example3.8 {
    Set time {
        Index: t;
    }
    Set customerload {
        Index: i;
    }
    Parameter energydemand {
        IndexDomain: (i);
    }
    Parameter TOUtarriff {
        IndexDomain: (i);
    }
    Parameter price {
        IndexDomain: (i);
    }
    Parameter no_timeslotrequired {
        IndexDomain: (i);
    }
    Parameter end_time {
        IndexDomain: (i);
    }
    Variable loadstatus {
        IndexDomain: (i,t);
        Range: binary;
    }
    Variable Obj_Function {
        Range: free;
        Definition: sum((t,i),(energydemand(i))*(loadstatus(i,t)*(price(i))));
    }
    Constraint constr1 {
        IndexDomain: (i);
        Definition: sum(t, loadstatus(i,t))=no_timeslotrequired(i);
        Comment: "sum(i, loadstatus(i,t))=no_timeslotrequired(i)";
    }
    Constraint constr2 {
        IndexDomain: (i);
        Definition: sum(t,(energydemand(i))* loadstatus(i,t))<=end_time(i);
    }
```

```
    MathematicalProgram Net_energy {
        Objective: Obj_Function;
        Direction: minimize;
        Constraints: AllConstraints;
        Variables: AllVariables;
        Type: Automatic;
    }
    Procedure MainInitialization {
        Comment: "Add initialization statements here that do NOT require any library
being initialized already.";
    }
    Procedure PostMainInitialization {
        Comment: {
            "Add initialization statements here that require that the libraries are
already initialized properly,
            or add statements that require the Data Management module to be
initialized."
        }
    }
    Procedure MainExecution {
        Body: {
            !solve NetCost;
        }
    }
```

### 3.7.2 *Incentive-Based Demand Response Modeling Using AIMMS*

The problem was introduced in Sect. 2.4. The *individual rationality constraint* and the *incentive compatibility constraint* are extended for a 24-hour period instead of a single time instant. In addition, budgetary limits and power constrains are incorporated. A complete description is provided in [14, 15]

**Set Definition**:

$t$ represents hourly time slots and $j$ represents the number of customers

**Variable Definition**:

$x_{j,t}$ is the amount of power either curtailed or shutdown by an electricity consumer.
$y_{j,t}$ is the incentive paid for power curtailed/shutdown.

**Parameter Definition**:

$\theta_j$ indicates the different kinds of consumers.
$\lambda_{j,t}$ is the locational impact of consumers at time t.
UB is the utility's total budget.

| | |
|---|---|
| $CM_j$ | is the daily limit of interruptible power for customer $j$. |
| $K_{1,\mathrm{j}}$ and $K_{2,\mathrm{j}}$ | are the cost function coefficients. |
| $J$ | is the total number of consumers. |
| $T$ | is is the total time interval. |

**Objective Function and Constraints**

$$\max_{x,y} \sum_{t=1}^{T} \sum_{j=1}^{J} \left[\lambda_{j,t} x_{j,t} - y_{j,t}\right] \tag{3.12}$$

s.t.

$$\sum_{t=1}^{T} \left[y_{j,t} - \left(K_{1,j} x_{j,t}^2 + K_{2,j} x_{j,t} - K_{2,t} x_{j,t} \theta_j\right)\right] \geq 0, \text{for} j = 1, \ldots, J. \tag{3.13}$$

$$\sum_{t=1}^{T} \left[y_{j,t} - \left(K_{1,j} x_{j,t}^2 + K_{2,j} x_{j,t} - K_{2,t} x_{j,t} \theta_j\right)\right], \geq \sum_{t=1}^{T} \left[y_{j-1,t} - \left(K_{1,j-1} x_{j-1,t}^2 + K_{2,j-1} x_{j-1,t} - K_{2,j-1} x_{j-1,t} \theta_{j-1}\right)\right]$$

$$\text{for} j = 1, \ldots, J. \tag{3.14}$$

$$\sum_{t=1}^{T} \sum_{j=1}^{J} y_{j,t} \leq \mathrm{UB} \tag{3.15}$$

$$\sum_{t=1}^{T} x_{j,t} \leq \mathrm{CM}_j, \quad \text{for} j = 1, \ldots, J. \tag{3.16}$$

where [14, 16]

Constraint (3.13) ensures that the total daily incentive received by a customer exceeds or equals his daily cost of interruption.
Constraint (3.14) ensures that the greater the customer power curtailed, the greater the customer benefit.
Constraint (3.15) ensures that the total incentive paid by the utility is less than the utility's budget.
Constraint (3.16) ensures that the total daily power curtailed by each customer is less than its daily limit of interruptible power.

## 3.8 Summary

This chapter covered the fundamentals of mathematical optimization, including classical, and meta-heuristic methods. Examples of prominent meta-heuristic and classical solvers were detailed. The nature of mathematical optimization and the steps to be taken from a modeler's perspective were also given. AMLs, specifically AIMMS, were introduced. Finally, examples were shows of different energy efficiency and DR programs and the notation in the AML format was given.

```
Model Example3.9 {
  Set time {
    Index: t, tt;
  }
  Set customer {
    Index: j;
  }
  Parameter totcustincentive {
    Definition: sum[ (t,j),y(j,t)];
  }
  Parameter totcustpower {
    Definition: sum[ (t,j),x(j,t)];
  }
  Parameter d {
    IndexDomain: t;
  }
  Parameter thet {
    IndexDomain: j;
  }
  Parameter k1 {
    IndexDomain: j;
  }
  Parameter k2 {
    IndexDomain: j;
  }
  Parameter lam {
    IndexDomain: (j,t);
  }
  Parameter lamp {
    IndexDomain: t;
  }
  Parameter w {
    IndexDomain: (t);
  }
  Parameter s {
    IndexDomain: t;
  }
  Parameter tcost {
    IndexDomain: j;
```

```
    Definition: sum[t,y(j,t)];
}
Parameter CM {
    IndexDomain: j;
}
Parameter tpow {
    IndexDomain: j;
    Definition: sum[t,x(j,t)];
}
Variable y {
    IndexDomain: (j,t);
    Range: nonnegative;
}
Variable uu {
    IndexDomain: (j,t);
    Range: free;
}
Variable x {
    IndexDomain: (j,t);
    Range: nonnegative;
}
Variable total {
    Range: free;
    Definition: sum[ (t,j), (lam(j,t)*x(j,t))-y(j,t)];
}
MathematicalProgram hmmm {
    Objective: total;
    Direction: maximize;
    Constraints: AllConstraints;
    Variables: AllVariables;
    Type: Automatic;
}
Constraint con1 {
    IndexDomain: (j,t);
    Definition: y(j,t)-((q1(j)*x(j,t)*x(j,t))+(q2(j)*x(j,t))-(q2(j)*x(g,t)*thet(j)))>=0;
}
Constraint con2 {
    IndexDomain: j;
```

```
    Definition: sum[t,(y(j,t)-((q1(j)*x(j,t)*x(j,t))+(q2(j)*x(j,t))-(q2(j)*x(j,t)*thet(j))))] >=sum[t,(z(j-1,t)-((q1(j-1)*x(j-1,t)*x(j-1,t))+(q2(j-1)*x(j-1,t))-(q2(j-1)*x(j-1,t)*thet(j-1))))];
  }
  Constraint con3 {
    Definition: sum[t,sum[g, y(j,t)]]<=10000;
  }
  Constraint con4 {
    IndexDomain: j;
    Definition: sum[t,x(j,t)]<=CM(j);
  }
  Procedure MainInitialization;
  Procedure MainExecution {
    Body: {
      solve hmmm;
      if(hmmm.ProgramStatus <> 'Optimal' ) then
      empty x,y,total;
      endif;
    }
  }
```

## References

1. García Vera Y, Dufo-López R, Bernal-Agustín J (2019) "Energy management in microgrids with renewable energy sources: A literature review", Appl Sci 9(18):3854
2. Renuka Kumari (2012) "Reliability of metallic framed structures", Ph.D. Thesis, Department of Civil Engineering, University of Kerala
3. Shehab M, Khader AT, Al-Betar MA (2017) A survey on applications and variants of the cuckoo search algorithm. Appl Soft Comput 61:1041–1059
4. Nazari-Heris M, Mohammadi-Ivatloo B, Gharehpetian GB (2018) A comprehensive review of heuristic optimization algorithms for optimal combined heat and power dispatch from economic and environmental perspectives. Renew Sustain Energy Rev 81(2):2128–2143
5. He S, Wu QH, Saunders JR (October 2009) "Group search optimizer: an optimization algorithm inspired by animal searching behaviour."IEEE Trans Evol Comput 13(5)
6. Gbadamosi SL, Nwulu NI (2020) "A multi-period composite generation and transmission expansion planning model incorporating renewable energy sources and demand response," Sustain Energy Technol Assessments 39(October 2019):100726
7. Gbadamosi SL, Nwulu NI (2020) "Optimal power dispatch and reliability analysis of hybrid CHP-PV-wind systems in farming applications," Sustainability 12
8. Fourer R (2013) "Algebraic modeling languages for optimization." In: Gass SI, Fu MC (eds) *Encyclopedia of operations research and management science*. Boston, MA:Springer
9. "INFORMS Impact Prize—INFORMS", Web.archive.org, 2020. [Online]. Available: https://web.archive.org/web/20131022091250/https://www.informs.org/Blogs/E-News-Blog/INFORMS-Impact-Prize. [Accessed: 10 Jul 2020]

10. AIMMS, “Download AIMMS developer”, 2020. [Online]. Available: https://www.aimms.com/english/developers/downloads/download-aimms/. [Accessed: 10 Jul 2020]
11. AIMMS,”Getting started”, 2020. [Online]. Available: https://www.aimms.com/english/developers/resources/getting-started/. [Accessed: 10 Jul 2020]
12. Gbadamosi SL, Nwulu NI, Sun Y (2018) Multi-objective optimization for composite generation and transmission expansion planning considering offshore wind power and feed in tariffs. IET Renew Power Gener 12(14):1687–1697
13. Nwulu N, Sun Y, “Optimal incentivized demand response and building retrofits”, 3rd International Conference on Advances in Computing, Communication and Engineering, ICACCE 2016, Durban, South Africa; 28 November 2016–29 November 2016
14. Nwulu NI, Xia X (2015) “A combined dynamic economic emission dispatch and time of use demand response mathematical modelling framework.”J Renew Sustain Energy, 4
15. Nwulu NI, Xia X (1 January 2015) “Multi-objective dynamic economic emission dispatch of electric power generation integrated with game theory based demand response programs.”Energy Convers Manage 89:963–974
16. Gbadamosi SL, Nwulu NI (2020) “Reliability assessment of composite generation and transmission expansion planning incorporating renewable energy sources,” J Renew Sustain Energy 026301(12)

# Chapter 4
# Dynamic Economic Emissions Dispatch

## 4.1 Introduction

This chapter describes the dynamic economic and emission dispatch (DEED) with the aim of determining the optimal power output of the generating units that will meet the load, while ensuring that the ramp and generating limits are not exceeded [1]. DEED is an essential optimization problem that gives the generation schedules of the forecast load demand through cost and emission minimization. The operating cost function can be modeled as a quadratic cost function while satisfying equality, inequality, and ramp rate constraints [2]. Here, we provide solutions for the DEED problems in AIMMS.

### *4.1.1 Dynamic Economic Dispatch*

Dynamic economic dispatch (DED) is an extension of static economic dispatch, which aids the generation dispatch of the available generating units. DED has the ability to forecast the load schedule to handle a sudden change in the power system demand [3]. DED has ramp rate constraints to maintain the generator's life and can forecast and dispatch load adequately, to avoid sudden changes in load demand. The optimization of the DED problem can be evaluated based on the power dispatched by the generating units.

In this section, the objective function of DED seeks to minimize the total operating cost to determine the dispatch level of the generating units, while satisfying the demand and other necessary constraints [4].

$$\text{Min} \sum_{k,i} C_i\left(P_{k,i}\right) \tag{4.1}$$



N. Nwulu and S. L. Gbadamosi, *Optimal Operation and Control of Power Systems Using an Algebraic Modelling Language*, Green Energy and Technology,
https://doi.org/10.1007/978-3-030-00395-1_4

$$C_i(P_{k,i}) = a_i + b_i P_{k,i} + c_i P_{k,i}^2 \tag{4.2}$$

subject to the following constraints:

$$\sum_i P_{k,i} = d_k + \text{loss}_k \tag{4.3}$$

$$\text{loss}_k = \sum_i \sum_j P_{k,i} bb_{i,j} P_{k,j} \tag{4.4}$$

$$P_i^{\min} \leq P_{k,i} \leq P_i^{\max} \tag{4.5}$$

$$P_{k,i} - P_{(k+1),i} \leq \text{DR}_i \tag{4.6}$$

$$P_{k,i} - P_{(k+1),i} \leq \text{UR}_i. \tag{4.7}$$

The desired variable result to be obtained by the optimization model is $P_{k,i}$.

Equation (4.2) is the fuel cost and it is assumed that it is a quadratic function of the power output of the generating units [5].

Constraint (4.3) is the supply–demand balance equation that states that the total power produced by the generating units equals the load demand and the total system losses [6]. Equation (4.4) is known as the B-coefficient method and it describes the best approach to calculate losses. Constraint (4.5) describes the generation limits and ensures that these limits do not exceed the prescribed limits. Constraints (4.6) and (4.7) are the generator ramp rates.

Table 4.1 presents the generator fuel cost coefficients as obtained in [1] and the hourly load profile is given in Fig. 4.1.

The optimization model is formulated and solved using the CPLEX solver embedded in AIMMS. The AIMMS code for solving an example of the DED problem is shown below: (Table 4.2)

**Table 4.1** Generator data

| Generator | $a_i$ | $b_i$ | $c_i$ | $P_i^{\min}$ | $P_i^{\max}$ | $R_i^{\text{down}}$ | $R_i^{\text{up}}$ |
|---|---|---|---|---|---|---|---|
| 1 | 240 | 7 | 0.007 | 100 | 500 | 120 | 80 |
| 2 | 200 | 10 | 0.0095 | 50 | 200 | 90 | 50 |
| 3 | 220 | 8.5 | 0.009 | 80 | 300 | 100 | 65 |
| 4 | 200 | 11 | 0.009 | 50 | 150 | 90 | 50 |
| 5 | 220 | 10.5 | 0.008 | 50 | 200 | 90 | 50 |
| 6 | 190 | 12 | 0.0075 | 50 | 120 | 90 | 50 |

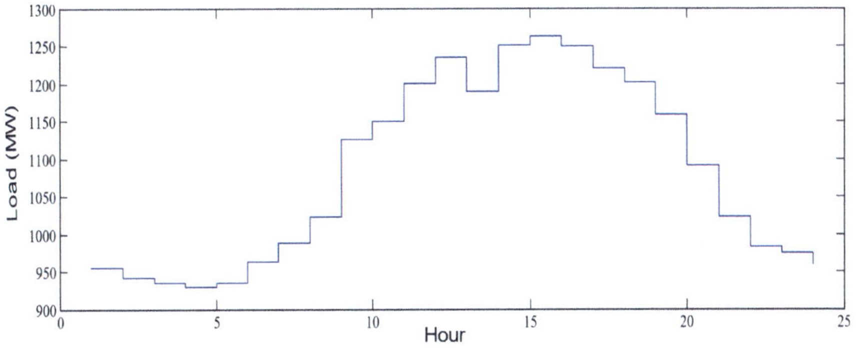


**Fig. 4.1** Total hourly load

**Table 4.2** Optimal generated power for DED model

| Time (hr) | Gen 1 | Gen 2 | Gen 3 | Gen 4 | Gen 5 | Gen 6 |
|---|---|---|---|---|---|---|
| 1 | 383.70 | 124.83 | 215.10 | 76.21 | 116.99 | 50 |
| 2 | 380.52 | 122.49 | 212.63 | 73.74 | 114.20 | 50 |
| 3 | 378.80 | 121.22 | 211.29 | 72.40 | 112.70 | 50 |
| 4 | 377.58 | 120.32 | 210.34 | 71.45 | 111.63 | 50 |
| 5 | 378.80 | 121.22 | 211.29 | 72.40 | 112.70 | 50 |
| 6 | 385.66 | 126.28 | 216.63 | 77.74 | 118.71 | 50 |
| 7 | 392.04 | 130.98 | 221.58 | 82.70 | 124.28 | 50 |
| 8 | 400.38 | 137.12 | 228.07 | 89.18 | 131.58 | 50 |
| 9 | 422.93 | 153.73 | 245.61 | 106.72 | 151.31 | 61.40 |
| 10 | 427.74 | 157.28 | 249.36 | 110.47 | 155.52 | 65.89 |
| 11 | 437.99 | 164.83 | 257.32 | 118.43 | 164.49 | 75.45 |
| 12 | 444.82 | 169.87 | 262.64 | 123.75 | 170.47 | 81.83 |
| 13 | 435.78 | 163.20 | 255.60 | 116.72 | 162.55 | 73.39 |
| 14 | 448.04 | 172.24 | 265.14 | 126.25 | 173.29 | 84.84 |
| 15 | 450.45 | 174.02 | 267.02 | 128.13 | 175.40 | 87.09 |
| 16 | 447.84 | 172.09 | 264.99 | 126.10 | 173.11 | 84.65 |
| 17 | 442.01 | 167.79 | 260.45 | 121.56 | 168.01 | 79.21 |
| 18 | 438.19 | 164.98 | 257.48 | 118.59 | 164.66 | 75.64 |
| 19 | 429.55 | 158.62 | 250.76 | 111.87 | 157.11 | 67.58 |
| 20 | 416.11 | 148.71 | 240.30 | 101.42 | 145.34 | 55.03 |
| 21 | 400.38 | 137.12 | 228.07 | 89.18 | 131.58 | 50 |
| 22 | 390.81 | 130.07 | 220.63 | 81.74 | 123.21 | 50 |
| 23 | 388.61 | 128.45 | 218.92 | 80.03 | 121.28 | 50.00 |
| 24 | 384.93 | 125.74 | 216.06 | 77.17 | 118.06 | 50.00 |

```
Model Main_Economic_Dispatch_bus_ded {
  Set Generators {
    Index: i, j;
  }
  Set Time {
    Index: k, kk, kkk;
  }
  Parameter pmin {
    IndexDomain: i;
  }
  Parameter pmax {
    IndexDomain: i;
  }
  Parameter a {
    IndexDomain: i;
  }
  Parameter e {
    IndexDomain: i;
  }
  Parameter f {
    IndexDomain: i;
  }
  Parameter g {
    IndexDomain: i;
  }
  Parameter y {
    Definition: 0;
  }
  Parameter z {
    Definition: 1-y;
  }
  Parameter b {
    IndexDomain: i;
  }
  Parameter c {
    IndexDomain: i;
  }
  Parameter ur {
    IndexDomain: i;
  }
  Parameter dr {
    IndexDomain: i;
  }
  Parameter d {
    IndexDomain: k;
  }
  Parameter bb {
    IndexDomain: (i,j);
  }
  Parameter loss {
    IndexDomain: k;
    Definition: sum[i,sum[j, p(k,i)*bb(i,j)*p(k,j)]];
  }
  Variable p {
    IndexDomain: (k,i);
    Range: nonnegative;
  }
  Variable Total {
    Range: free;
    Definition: sum[ (k,i), a(i)+(b(i)*p(k,i))+(c(i)*p(k,i)*p(k,i)) ];
  }
  Parameter tloss {
    Definition: sum[k,loss(k)];
  }
  Parameter costt {
    Definition: (sum[ (k,i), (a(i)+(b(i)*p(k,i))+(c(i)*p(k,i)*p(k,i)))]);
  }
```

```
    Parameter emissions {
        Definition: (sum[ (k,i), e(i)+(f(i)*p(k,i))+(g(i)*p(k,i)*p(k,i))]);
    }
    Parameter totgenpower {
        Definition: sum[ (i,k),p(k,i)];
    }
    Constraint sup {
        IndexDomain: k;
        Definition: sum[i, p(k,i)]=d(k)+loss(k);
    }
    MathematicalProgram LeastEconDis {
        Objective: Total;
        Direction: minimize;
        Constraints: AllConstraints;
        Variables: AllVariables;
        Type: Automatic;
    }
    Constraint sup1 {
        IndexDomain: (k,i);
        Definition: p(k,i)<=pmax(i);
    }
    Constraint sup2 {
        IndexDomain: (k,i);
        Definition: p(k,i)>=pmin(i);
    }
    Constraint sup3 {
        IndexDomain: (kk,i) | 1<= ord(kk) <=23;
        Definition: p(kk,i)-p(kk+1,i)<=dr(i);
    }
    Constraint sup4 {
        IndexDomain: (kk,i) | 1<= ord(kk) <=23;
        Definition: p(kk+1,i)-p(kk,i)<=ur(i);
    }
    Procedure MainInitialization;
    Procedure MainExecution {
        Body: {
            solve LeastEconDis;
        }
    }
    Procedure MainTermination {
        Body: {
            return DataManagementExit();
        }
    }
}
```

Figure 4.2 presents the results obtainable from the optimization and the graphical representation is presented in Fig. 4.2. The simulated results show variation in the optimal power output from the generators owing to different fuel costs for each generator. The generator with minimal cost has a larger power output, while the generator with higher cost has lower power output.

### 4.1.2 Dynamic Economic Emissions Dispatch

DEED is a strategy for integrating emission dispatch into the DED optimization problem. DEED is an essential optimization problem in power system operation to satisfy both economic and emission perspectives [7, 8]. DEED's objective is to minimize the cost and emissions while satisfying technical and economic

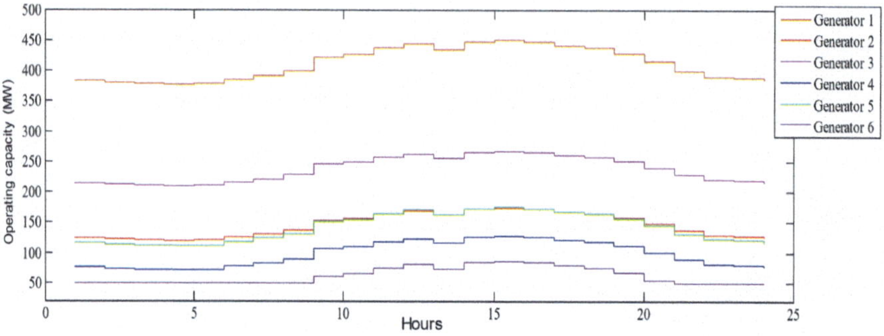


**Fig. 4.2** Optimal capacity of generating units

constraints with a view to obtaining the required generator power outputs over a specific period [9].

$$\text{Min}\left[y\sum_{k,i} C_i\left(P_{k,i}\right) + z\sum_{k,i} E_i\left(P_{k,i}\right)\right] \tag{4.8}$$

$$C_i(P_{k,i}) = a_i + b_i P_{k,i} + c_i P_{k,i}^2 \tag{4.9}$$

$$E_i(P_{k,i}) = e_i + f_i P_{k,i} + g_i P_{k,i}^2 \tag{4.10}$$

$$y + z = 1 \tag{4.11}$$

subject to the following constraints:

$$\sum_i P_{k,i} = d_k + \text{loss}_k \tag{4.12}$$

$$\text{loss}_k = \sum_i \sum_j P_{k,i} \text{bb}_{i,j} P_{k,j} \tag{4.13}$$

$$P_i^{\min} \le P_{k,i} \le P_i^{\max} \tag{4.14}$$

$$P_{k,i} - P_{(k+1),i} \le R_i^{\text{down}} \tag{4.15}$$

$$P_{k,i} - P_{(k+1),i} \le R_i^{\text{up}}. \tag{4.16}$$

The desired variable result to be obtained by the optimization model is $P_{k,i}$.

For simplicity of the optimization problem, a quadratic cost function is applied for the fuel and emission cost functions, as presented in Eqs. (4.9) and (4.10) [10]. Equation (4.8) describes the DEED as having a multi-objective optimization problem, namely, two conflicting objectives (economic and emission). Therefore, to convert this into a single-objective, a weighting factor approach is applied, as presented in Eq. (4.11) [11]. Here, equal preference is given to the two objectives and the optimization is done simultaneously [12].

Constraint (4.12) is the supply–demand balance equation that states that the total power produced by the generating units equals the load demand and the total system losses. Equation (4.13) is known as the B-coefficient method and describes the best approach to calculate losses. Constraint (4.14) describes the generation limits and ensures that these limits do not exceed the stipulated limit. Constraints (4.15) and (4.16) are the generator ramp rates.

Table 4.3 presents the generator fuel cost coefficients as obtained in [1] and the hourly load profile is given in Fig. 4.3. The optimization model is formulated and solved using the CPLEX solver embedded in AIMMS. The AIMMS code for solving an example of a DEED problem is shown below.

**Table 4.3** Generator data

| Generator | $a_i$ | $b_i$ | $c_i$ | $e_i$ | $f_i$ | $g_i$ | $P_i^{min}$ | $P_i^{max}$ | $R_i^{down}$ | $R_i^{up}$ |
|---|---|---|---|---|---|---|---|---|---|---|
| 1 | 240 | 7 | 0.007 | 13.8593 | 0.32767 | 0.00419 | 100 | 500 | 120 | 80 |
| 2 | 200 | 10 | 0.0095 | 13.8593 | 0.32767 | 0.00419 | 50 | 200 | 90 | 50 |
| 3 | 220 | 8.5 | 0.009 | 40.2669 | −0.54551 | 0.00683 | 80 | 300 | 100 | 65 |
| 4 | 200 | 11 | 0.009 | 40.2669 | −0.54551 | 0.00683 | 50 | 150 | 90 | 50 |
| 5 | 220 | 10.5 | 0.008 | 42.8955 | −0.51116 | 0.00461 | 50 | 200 | 90 | 50 |
| 6 | 190 | 12 | 0.0075 | 42.8955 | −0.51116 | 0.00461 | 50 | 120 | 90 | 50 |

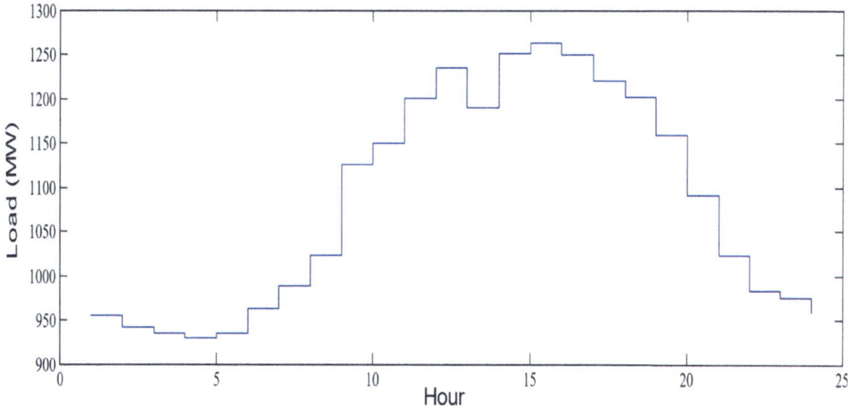


**Fig. 4.3** Total hourly load

The AIMMS code for solving the DEED is given below:

```
Model Main_Economic_Emission Dispatch_bus_deed {
  Set Generators {
    Index: i, j;
  }
  Set Time {
    Index: k, kk, kkk;
  }
  Parameter pmin {
    IndexDomain: i;
  }
  Parameter pmax {
    IndexDomain: i;
  }
  Parameter a {
    IndexDomain: i;
  }
  Parameter e {
    IndexDomain: i;
  }
  Parameter f {
    IndexDomain: i;
  }
  Parameter g {
    IndexDomain: i;
  }
  Parameter y {
    Definition: 0;
  }
  Parameter z {
    Definition: 1-y;
  }
  Parameter b {
    IndexDomain: i;
  }
  Parameter c {
    IndexDomain: i;
  }
  Parameter ur {
    IndexDomain: i;
  }
  Parameter dr {
    IndexDomain: i;
  }
  Parameter d {
    IndexDomain: k;
  }
  Parameter bb {
    IndexDomain: (i,j);
  }
  Parameter loss {
    IndexDomain: k;
    Definition: sum[i,sum[j, p(k,i)*bb(i,j)*p(k,j)]];
  }
  Variable p {
    IndexDomain: (k,i);
    Range: nonnegative;
  }
  Variable Total {
    Range: free;
    Definition: (y*sum[ (k,i), a(i)+(b(i)*p(k,i))+(c(i)*p(k,i)*p(k,i)) ])+(z*sum[ (k,i), e(i)+(f(i)*p(k,i))+(g(i)*p(k,i)*p(k,i)) ]);
  }
  Parameter tloss {
    Definition: sum[k,loss(k)];
  }
  Parameter costt {
    Definition: (sum[ (k,i), (a(i)+(b(i)*p(k,i))+(c(i)*p(k,i)*p(k,i)))]);
  }
```

```
Parameter emissions {
   Definition: (sum[ (k,i), e(i)+(f(i)*p(k,i))+(g(i)*p(k,i)*p(k,i))]);
}
Parameter totgenpower {
   Definition: sum[ (i,k),p(k,i)];
}
Constraint sup {
   IndexDomain: k;
   Definition: sum[i, p(k,i)]=d(k)+loss(k);
}
MathematicalProgram LeastEconDis {
   Objective: Total;
   Direction: minimize;
   Constraints: AllConstraints;
   Variables: AllVariables;
   Type: Automatic;
}
Constraint sup1 {
   IndexDomain: (k,i);
   Definition: p(k,i)<=pmax(i);
}
Constraint sup2 {
   IndexDomain: (k,i);
   Definition: p(k,i)>=pmin(i);
}
Constraint sup3 {
   IndexDomain: (kk,i) | 1<= ord(kk) <=23;
   Definition: p(kk,i)-p(kk+1,i)<=dr(i);
}
Constraint sup4 {
   IndexDomain: (kk,i) | 1<= ord(kk) <=23;
   Definition: p(kk+1,i)-p(kk,i)<=ur(i);
}
Procedure MainInitialization;
Procedure MainExecution {
   Body: {
      solve LeastEconDis;
   }
}
Procedure MainTermination {
   Body: {
      return DataManagementExit();
   }
}
}
```

Table 4.4 presents the results obtainable from the optimization. Compared to the DED model, the simulated results obtained in the DEED model show that the optimal power output for generators 1, 2, 3, and 4 decreased owing to high emissions from the generating units involved, while the power output for generators 5 and 6 increased owing to low generator emissions. Therefore, generating units with lower emission quantities give better performance than generating units with higher emission quantities.

## 4.2 Emission-Constrained Dynamic Economic Dispatch

To optimize the operating cost and emission adequately, economic and emission scheduling involves sharing of the generation among the available generating units while satisfying the constraints [13]. There is growing concern about the emissions

**Table 4.4** Optimal generated power for DEED model

| Time (hr) | Gen 1 | Gen 2 | Gen 3 | Gen 4 | Gen 5 | Gen 6 |
|---|---|---|---|---|---|---|
| 1 | 164.41 | 164.41 | 164.78 | 150 | 200 | 120 |
| 2 | 159.35 | 159.35 | 161.68 | 150 | 200 | 120 |
| 3 | 156.62 | 156.62 | 160.01 | 150 | 200 | 120 |
| 4 | 154.68 | 154.68 | 158.81 | 150 | 200 | 120 |
| 5 | 156.62 | 156.62 | 160.01 | 150 | 200 | 120 |
| 6 | 167.53 | 167.53 | 166.69 | 150 | 200 | 120 |
| 7 | 177.66 | 177.66 | 172.91 | 150 | 200 | 120 |
| 8 | 190.92 | 190.92 | 181.05 | 150 | 200 | 120 |
| 9 | 250.67 | 200 | 217.70 | 150 | 200 | 120 |
| 10 | 265.96 | 200 | 227.08 | 150 | 200 | 120 |
| 11 | 298.49 | 200 | 247.04 | 150 | 200 | 120 |
| 12 | 320.23 | 200 | 260.37 | 150 | 200 | 120 |
| 13 | 291.47 | 200 | 242.73 | 150 | 200 | 120 |
| 14 | 330.47 | 200 | 266.65 | 150 | 200 | 120 |
| 15 | 338.15 | 200 | 271.37 | 150 | 200 | 120 |
| 16 | 329.83 | 200 | 266.26 | 150 | 200 | 120 |
| 17 | 311.27 | 200 | 254.88 | 150 | 200 | 120 |
| 18 | 299.13 | 200 | 247.43 | 150 | 200 | 120 |
| 19 | 271.69 | 200 | 230.60 | 150 | 200 | 120 |
| 20 | 229.04 | 200 | 204.43 | 150 | 200 | 120 |
| 21 | 190.92 | 190.92 | 181.05 | 150 | 200 | 120 |
| 22 | 175.71 | 175.71 | 171.72 | 150 | 200 | 120 |
| 23 | 172.20 | 172.20 | 169.56 | 150 | 200 | 120 |
| 24 | 166.36 | 166.36 | 165.98 | 150 | 200 | 120 |

contribution from generating units [14]. This has enforced additional constraint on the DED because of environmental considerations, thus affecting the generation schedule. The emission-constrained economic dispatch of the thermal generators is an optimization problem that seeks to minimize the operating cost while satisfying the technical and economic constraints as well as the emission constraint.

$$\text{Min} \sum_{k,i} C_i\left(P_{k,i}\right) \tag{4.17}$$

$$C_i(P_{k,i}) = a_i + b_i P_{k,i} + c_i P_{k,i}^2 \tag{4.18}$$

subject to the following constraints:

$$\sum_i P_{k,i} = d_k + \text{loss}_k \tag{4.19}$$

$$\text{loss}_k = \sum_i \sum_j P_{k,i} \text{bb}_{i,j} P_{k,j} \tag{4.20}$$

$$P_i^{\min} \le P_{k,i} \le P_i^{\max} \tag{4.21}$$

$$P_{k,i} - P_{(k+1),i} \le R_i^{\text{down}} \tag{4.22}$$

$$P_{k,i} - P_{(k+1),i} \le R_i^{\text{up}} \tag{4.23}$$

$$\sum_{k,i} H_i(P_{k,i}) \le E^{\max}. \tag{4.24}$$

The desired variable result to be obtained by the optimization model is $P_{k,i}$.

The objective function is presented in Eq. (4.17) and its quadratic cost function is given in Eq. (4.18). Constraint (4.19) is the supply–demand balance equation that states that the total power produced by the generating units equals the load demand and the total system losses. Equation (4.20) is known as the B-coefficient method and it describes the best approach to calculate losses. Constraint (4.21) describes the generation limits and it ensures that these limits do not exceed the stipulated limit. Constraints (4.22) and (4.23) are the generator ramp rates. Constraint (4.24) enforces that the sum of emissions from the generating units must be less than or equal to the maximum permissible emissions. Table 4.5 presents the generator fuel cost coefficients and emission quantities for a 30-bus system with six generators. These data are obtained in [15] and the hourly load profile is given in Fig. 4.4.

**Table 4.5** Generator data for emission-constrained DED model

| Generator | $a_i$ | $b_i$ | $c_i$ | $e_i$ | $f_i$ | $g_i$ | $P_i^{\min}$ | $P_i^{\max}$ | $R_i^{\text{down}}$ | $R_i^{\text{up}}$ | $H_i$ |
|---|---|---|---|---|---|---|---|---|---|---|---|
| 1 | 240 | 7 | 0.007 | 13.8593 | 0.32767 | 0.00419 | 100 | 500 | 120 | 80 | 0.5 |
| 2 | 200 | 10 | 0.0095 | 13.8593 | 0.32767 | 0.00419 | 50 | 200 | 90 | 50 | 0.5 |
| 3 | 220 | 8.5 | 0.009 | 40.2669 | −0.54551 | 0.00683 | 80 | 300 | 100 | 65 | 0.5 |
| 4 | 200 | 11 | 0.009 | 40.2669 | −0.54551 | 0.00683 | 50 | 150 | 90 | 50 | 0.5 |
| 5 | 220 | 10.5 | 0.008 | 42.8955 | −0.51116 | 0.00461 | 50 | 200 | 90 | 50 | 0.5 |
| 6 | 190 | 12 | 0.0075 | 42.8955 | −0.51116 | 0.00461 | 50 | 120 | 90 | 50 | 0.5 |

The AIMMS code for solving the emission-constrained DED is given below:

```
Model Main_Emission Constrained_ded {
  Set Generators {
    Index: i, j;
  }
  Set Time {
    Index: k, kk, kkk;
  }
  Parameter pmin {
    IndexDomain: i;
  }
  Parameter pmax {
    IndexDomain: i;
  }
  Parameter a {
    IndexDomain: i;
  }
  Parameter e {
    IndexDomain: i;
  }
  Parameter f {
    IndexDomain: i;
  }
  Parameter g {
    IndexDomain: i;
  }
  Parameter y {
    Definition: 0;
  }
  Parameter z {
    Definition: 1-y;
  }
  Parameter b {
    IndexDomain: i;
  }
  Parameter c {
    IndexDomain: i;
  }
  Parameter ur {
    IndexDomain: i;
  }
  Parameter dr {
    IndexDomain: i;
  }
  Parameter d {
    IndexDomain: k;
  }
  Parameter bb {
    IndexDomain: (i,j);
  }
  Parameter loss {
    IndexDomain: k;
    Definition: sum[i,sum[j, p(k,i)*bb(i,j)*p(k,j)]];
  }
  Parameter H {
    IndexDomain: i;
  }
  Parameter Emax;
  Variable p {
    IndexDomain: (k,i);
    Range: nonnegative;
```

```
    }
    Variable Total {
        Range: free;
        Definition: sum[ (k,i), a(i)+(b(i)*p(k,i))+(c(i)*p(k,i)*p(k,i)) ];
    }
    Parameter tloss {
        Definition: sum[k,loss(k)];
    }
    Parameter costt {
        Definition: (sum[ (k,i), (a(i)+(b(i)*p(k,i))+(c(i)*p(k,i)*p(k,i)))]);
    }
    Parameter emissions {
        Definition: (sum[ (k,i), e(i)+(f(i)*p(k,i))+(g(i)*p(k,i)*p(k,i))]);
    }
    Parameter totgenpower {
        Definition: sum[ (i,k),p(k,i)];
    }
    Constraint sup {
        IndexDomain: k;
        Definition: sum[i, p(k,i)]=d(k)+loss(k);
    }
    MathematicalProgram LeastEconDis {
        Objective: Total;
        Direction: minimize;
        Constraints: AllConstraints;
        Variables: AllVariables;
        Type: Automatic;
    }
    Constraint sup1 {
        IndexDomain: (k,i);
        Definition: p(k,i)<=pmax(i);
    }
    Constraint sup2 {
        IndexDomain: (k,i);
        Definition: p(k,i)>=pmin(i);
    }
    Constraint sup3 {
        IndexDomain: (kk,i) | 1<= ord(kk) <=23;
        Definition: p(kk,i)-p(kk+1,i)<=dr(i);
    }
    Constraint sup4 {
        IndexDomain: (kk,i) | 1<= ord(kk) <=23;
        Definition: p(kk+1,i)-p(kk,i)<=ur(i);
    }
    Constraint sup5 {
        Definition: sum[ (k,i), (H(i)*p(k,i)) ]<=Emax;
    }
    Procedure MainInitialization;
    Procedure MainExecution {
        Body: {
            solve LeastEconDis;
        }
}
```

Table 4.6 presents the optimal power output for the six generators. From the simulated results, it can be observed that the generators with higher operational costs give lower power output, while the opposite happens for generators with lower operational costs.

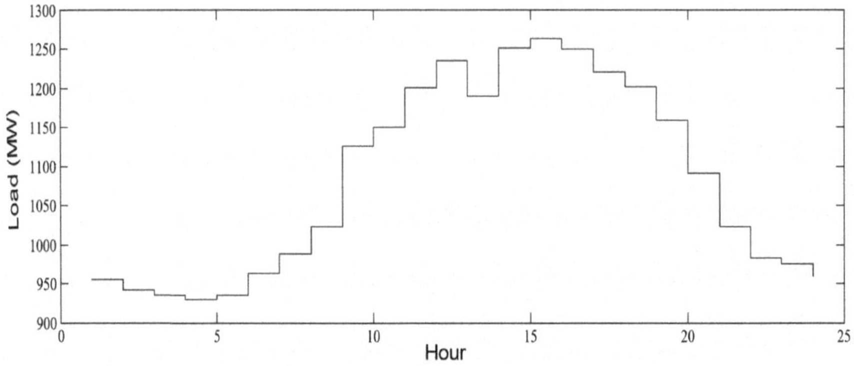


**Fig. 4.4** Total hourly load for emission-constrained DED model

**Table 4.6** Optimal generated power for emission-constrained DED model

| Time (hr) | Gen 1 | Gen 2 | Gen 3 | Gen 4 | Gen 5 | Gen 6 |
|---|---|---|---|---|---|---|
| 1 | 383.70 | 124.83 | 215.10 | 76.21 | 116.99 | 50 |
| 2 | 380.52 | 122.49 | 212.63 | 73.74 | 114.20 | 50 |
| 3 | 378.80 | 121.22 | 211.29 | 72.40 | 112.70 | 50 |
| 4 | 377.58 | 120.32 | 210.34 | 71.45 | 111.63 | 50 |
| 5 | 378.80 | 121.22 | 211.29 | 72.40 | 112.70 | 50 |
| 6 | 385.66 | 126.28 | 216.63 | 77.74 | 118.71 | 50 |
| 7 | 392.04 | 130.98 | 221.58 | 82.70 | 124.28 | 50 |
| 8 | 400.38 | 137.12 | 228.07 | 89.18 | 131.58 | 50 |
| 9 | 422.93 | 153.73 | 245.61 | 106.72 | 151.31 | 61.40 |
| 10 | 427.74 | 157.28 | 249.36 | 110.47 | 155.52 | 65.89 |
| 11 | 437.99 | 164.83 | 257.32 | 118.43 | 164.49 | 75.45 |
| 12 | 444.82 | 169.87 | 262.64 | 123.75 | 170.47 | 81.83 |
| 13 | 435.78 | 163.20 | 255.60 | 116.72 | 162.55 | 73.39 |
| 14 | 448.04 | 172.24 | 265.14 | 126.25 | 173.29 | 84.84 |
| 15 | 450.45 | 174.02 | 267.02 | 128.13 | 175.40 | 87.09 |
| 16 | 447.84 | 172.09 | 264.99 | 126.10 | 173.11 | 84.65 |
| 17 | 442.01 | 167.79 | 260.45 | 121.56 | 168.01 | 79.21 |
| 18 | 438.19 | 164.98 | 257.48 | 118.59 | 164.66 | 75.64 |
| 19 | 429.55 | 158.62 | 250.76 | 111.87 | 157.11 | 67.58 |
| 20 | 416.11 | 148.71 | 240.30 | 101.42 | 145.34 | 55.03 |
| 21 | 400.38 | 137.12 | 228.07 | 89.18 | 131.58 | 50 |
| 22 | 390.81 | 130.07 | 220.63 | 81.74 | 123.21 | 50 |
| 23 | 388.61 | 128.45 | 218.92 | 80.03 | 121.28 | 50 |
| 24 | 384.93 | 125.74 | 216.06 | 77.17 | 118.06 | 50 |

## 4.3 Bid-Based Dynamic Economic Emission Dispatch

The problem associated with determination of power dispatch of the generating units with the objective of reducing emissions as well as minimizing operational costs has been reviewed by different scholars in the past. The idea of power system deregulation leads to the unbundling of the grid into generation, transmission and distribution, crafting an avenue for the electricity market. Under these privatization schemes, there are two market strategies, namely, bid-based DEED and price-based DEED. In the bid-based DEED approach, an independent system operator (ISO) supervises the electricity trading system. The ISO aims to maximize the social profit by minimizing the difference between operational cost and customer bids, as well as the emissions. The bid-based DEED gives the generators and customers bid quantities over a specified operational scheduling. The mathematical formulation is given as:

$$\text{Min}\left[w_1\left\{\sum_{k=1}\sum_{i=1} C_i\left(P_{k,i}\right) - \sum_{k=1}\sum_{y=1} \delta_g\left(B_{k,g}\right)\right\} + w_2\sum_{k=1}\sum_{i=1} E_i\left(P_{k,i}\right)\right] \tag{4.25}$$

$$C_i(P_{k,i}) = a_i + b_i P_{k,i} + c_i P_{k,i}^2 \tag{4.26}$$

$$\delta_g\left(B_{k,g}\right) = m_g B_{k,g} + n_g B_{k,g}^2 \tag{4.27}$$

subject to the following constraints:

$$\sum_i P_{k,i} = \sum_{g=1} B_g + \text{loss}_k \tag{4.28}$$

$$\text{loss}_k = \sum_i \sum_j P_{k,i} bb_{i,j} P_{k,j} \tag{4.29}$$

$$B_g^{\min} \leq B_{k,g} \leq B_g^{\max} \tag{4.30}$$

$$P_i^{\min} \leq P_{k,i} \leq P_i^{\max} \tag{4.31}$$

$$P_{k,i} - P_{(k+1),i} \leq \text{DR}_i \tag{4.32}$$

$$P_{k,i} - P_{(k+1),i} \leq \text{UR}_i. \tag{4.33}$$

The desired variable result to be obtained by the optimization model is $P_{k,i}$ and $B_{k,g}$.

The quadratic cost functions for the fuel and customers' bid quantities are presented in Eqs. (4.26) and (4.27). Constraint (4.28) is the supply–demand balance equation that states that the sum of generator bid quantities must be equal to the customer bid quantities and the system losses. Equation (4.29) is known as the B-coefficient method and it describes the best approach to calculate losses. Constraint (30) ensures that the bid quantities by the customer are within the limits. Constraint (4.31) describes the generation limits and it ensures that these limits do not exceed the stipulated limit. Constraints (4.32) and (4.33) are the generator ramp rates.

The bid-based DEED optimization problem is applied to the IEEE 30-bus system, as shown in Fig. 4.5. The test bus system consists of six generators and two customers. Tables 4.7 and 4.8 present the bid data for the generators and customers, respectively. The coefficient for B loss is depicted by Eq. (4.34).

The optimization model is formulated and solved using the CPLEX solver embedded in AIMMS.

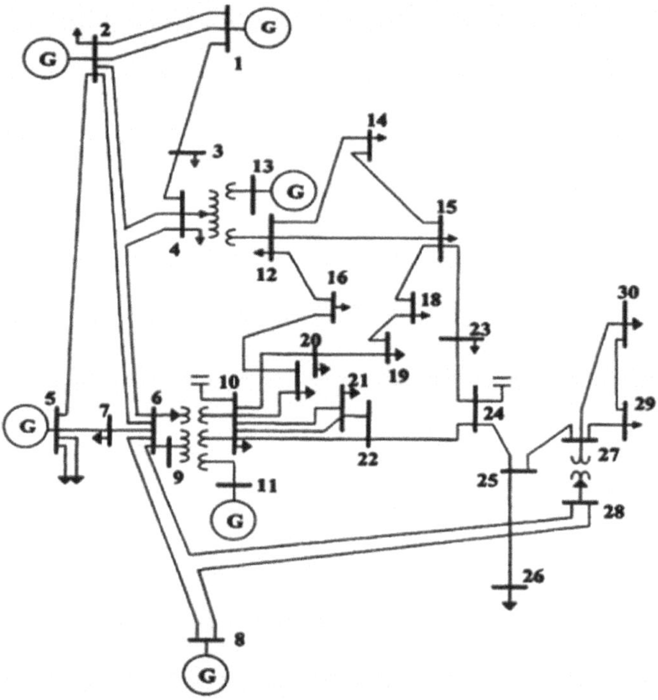


**Fig. 4.5** IEEE 30-bus system

**Table 4.7** Generator data for bid-based DEED model

| Generator | $a_i$(\$/h) | $b_i$(\$/MWh) | $c_i$(\$/MW$^2$h) | $P_i^{min}$ (MW) | $P_i^{max}$ (MW) | DR$_i$ (MW/h) | UR$_i$(MW/h) |
|---|---|---|---|---|---|---|---|
| 1 | 0.00375 | 2 | 0 | 50 | 200 | 85 | 65 |
| 2 | 0.0175 | 1.75 | 0 | 20 | 80 | 22 | 12 |
| 3 | 0.00625 | 1 | 0 | 15 | 50 | 15 | 12 |
| 4 | 0.00834 | 3.25 | 0 | 10 | 35 | 16 | 8 |
| 5 | 0.025 | 3 | 0 | 10 | 30 | 9 | 6 |
| 6 | 0.025 | 3 | 0 | 12 | 40 | 16 | 8 |

The AIMMS code for solving the bid-based DEED is given below:

```
Model Main_Emission_constrained_deed {
  Set time {
    Index: t, tt;
  }
  Set customer {
    Index: j;
  }
  Set generator {
    Index: i, k;
  }
  Parameter pmin {
    IndexDomain: i;
  }
  Parameter pmax {
    IndexDomain: i;
  }
  Parameter xmin {
    IndexDomain: (j,t);
  }
  Parameter xmax {
    IndexDomain: (j,t);
  }
  Parameter a {
    IndexDomain: i;
  }
  Parameter b {
    IndexDomain: i;
  }
  Parameter c {
    IndexDomain: i;
  }
  Parameter e {
    IndexDomain: j;
  }
  Parameter f {
    IndexDomain: j;
  }
  Parameter ur {
    IndexDomain: i;
  }
  Parameter dr {
    IndexDomain: i;
  }
  Parameter bb {
    IndexDomain: (i,k);
  }
  Parameter loss {
    IndexDomain: t;
    Definition: sum[i,sum[k, p(i,t)*bb(i,k)*p(k,t)]];
  }
  Parameter tloss {
    Definition: sum[t,loss(t)];
  }
  Parameter tgencost {
    Definition: (sum[ (t,i), c(i)+(b(i)*p(i,t))+(a(i)*p(i,t)*p(i,t))]);
  }
  Parameter tbidcost {
    Definition: sum[ (t,j), (f(j)*x(j,t))+(e(j)*x(j,t)*x(j,t)) ];
  }
  Parameter totcustpower {
    Definition: sum[ (t,j),x(j,t)];
  }
  Parameter totpowgen {
    Definition: sum[ (i,t),p(i,t)];
  }
  Parameter thet {
    IndexDomain: i;
```

```
}
Parameter totalloss {
   Definition: sum[t,loss(t)];
}
Parameter totalemissions {
   Definition: sum[t,sum[i, thet(i)* p(i,t)]];
}
Parameter profit {
   Definition: (sum[ (t,j), (f(j)*x(j,t))+(e(j)*x(j,t)*x(j,t)) ])-(sum[ (t,i), c(i)+(b(i)*p(i,t))+(a(i)*p(i,t)*p(i,t))]);
}
Variable x {
   IndexDomain: (j,t);
   Range: free;
}
Variable p {
   IndexDomain: (i,t);
   Range: nonnegative;
}
Variable total {
   Range: free;
   Definition: (sum[ (t,j), (f(j)*x(j,t))+(e(j)*x(j,t)*x(j,t)) ])-(sum[ (t,i), c(i)+(b(i)*p(i,t))+(a(i)*p(i,t)*p(i,t))]);
}
MathematicalProgram hmmm {
   Objective: total;
   Direction: maximize;
   Constraints: AllConstraints;
   Variables: AllVariables;
   Type: Automatic;
}
Constraint con1 {
   IndexDomain: (j,t);
   Definition: x(j,t)>=xmin(j,t);
}
Constraint con4 {
   IndexDomain: t;
   Definition: sum[i, p(i,t)]=sum[j, x(j,t)]+loss(t);
}
Constraint con5 {
   IndexDomain: (i,t);
   Definition: p(i,t)<=pmax(i);
}
Constraint con6 {
   IndexDomain: (i,t);
   Definition: p(i,t)>=pmin(i);
}
Constraint con7 {
   IndexDomain: (i,t) | 1<= ord(t) <=2;
   Definition: p(i,t)-p(i,t+1)<=dr(i);
}
Constraint con8 {
   IndexDomain: (i,t) | 1<= ord(t) <=2;
   Definition: p(i,t+1)-p(i,t)<=ur(i);
}
Constraint con9 {
   Definition: sum[t,sum[i, thet(i)* p(i,t)]]<=3300;
}
Constraint con10 {
   IndexDomain: (j,t);
   Definition: x(j,t)<=xmax(j,t);
}
Procedure MainInitialization;
Procedure MainExecution {
   Body: {
      solve hmmm;
   }
```

**Table 4.8** Data for customer bid

| Customer | $m_g$($/MWh) | $n_g$($/MW2h) | $B_g^{min}$ (MW) | $B_g^{max}$(MW) |
|---|---|---|---|---|
| 1 | 20 | −0.06 (Low)<br>0.07Medium)<br>0.1 (High) | (t = 1) 100<br>(t = 2) 20 | (t = 1) 150<br>(t = 2) 70 |
| 2 | 15 | −0.08 (Low)<br>0.05(Medium)<br>0.09 (High) | (t = 1) 50<br>(t = 2) 100 | (t = 1) 100<br>(t = 2) 200 |

$$B = 10^{-4}\begin{bmatrix} 0.420 & 0.051 & 0.045 & 0.057 & 0.078 & 0.066 \\ 0.051 & 0.180 & 0.039 & 0.048 & 0.045 & 0.060 \\ 0.045 & 0.039 & 0.195 & 0.051 & 0.072 & 0.057 \\ 0.057 & 0.048 & 0.051 & 0.213 & 0.090 & 0.075 \\ 0.078 & 0.045 & 0.072 & 0.090 & 0.207 & 0.096 \\ 0.066 & 0.060 & 0.057 & 0.075 & 0.096 & 0.255 \end{bmatrix} \text{per MW} \quad (4.34)$$

Sensitivity analysis was carried out on the IEEE 30-bus test system using three customer bids. These bids are assumed to be low, medium, and high customer bids, as indicated in Table 4.9. The table presents the optimal results for the medium customer bid and the overall results for the three customer bids are given in the table. The results show that the lower customer bid quantity gives a large profit compared to high customer bids with a decreased profit (Table 4.10).

**Table 4.9** Results for optimal bids

| Medium Bid | | |
|---|---|---|
| Customer | t = 1 | t = 2 |
| 1 | 141.8548 | 62.10805 |
| 2 | 75.77397 | 100.1205 |
| Generators | | |
| 1 | 122.5859 | 84.88417 |
| 2 | 33.61911 | 21.91982 |
| 3 | 29.64141 | 15 |
| 4 | 10.17819 | 15.92708 |
| 5 | 10.31158 | 9 |
| 6 | 12.13455 | 15.93043 |

**Table 4.10** Generator and customer bid costs for IEEE 30-bus system

| IEEE Test System | 30-Bus (Low Bid) | 30-Bus (Medium Bid) | 30-Bus (High Bid) |
|---|---|---|---|
| Total generator bid cost ($) | 919.89246 | 919.89246 | 1017.4145 |
| Total customer bid cost ($) | 4017.5978 | 9184.5798 | 11669.457 |
| Total profit ($) | 3098 | 8264.6873 | 10652.043 |
| Total generator bid (MW) | 381.13221 | 381.13221 | 413.57777 |
| Total customer bid (MW) | 379.85736 | 379.85736 | 411.94749 |
| Total losses (MW) | 1.2748544 | 1.2748544 | 1.6302794 |
| Total emissions (MW) | 190.56611 | 190.56611 | 206.78888 |

## 4.4 Incentive-Based Dynamic Economic Emissions Dispatch

This section investigates the integration of dynamic economic emission dispatch with a DR program. This is absolutely essential for proper power system management in view of the ability of DR programs to control and curtail energy demand at the customer side [1]. Therefore, integrating both optimization problems would yield a better practical power system with reasonable solutions on both the generation and demand side. This implies inclusion of incentive-based DR into DEED such that customers benefit from the incentive offered for participating in DR [9]. The main objective is to determine the optimal power dispatch while considering minimizing fuel and emission costs, as well as maximizing the DR benefits for the customer. The mathematical formulation is presented below:

$$\text{Min}\left[w_1\left(\sum_{k=1}^{K}\sum_{i=1}^{I}C_i(P_{i,k})\right)+w_2\left(\sum_{k=1}^{K}\sum_{i=1}^{I}E_i(P_{i,k})\right)-w_3\left(\sum_{k=1}^{K}\sum_{g=1}^{G}(\lambda_{g,k}x_{g,k}-z_{g,k})\right)\right] \tag{4.35}$$

$$C_i(P_{i,k})=a_i+b_iP_{i,k}+c_iP_{i,k}^2 \tag{4.36}$$

$$E_i(P_{i,k}) = e_i + f_i P_{i,k} + g_i P_{i,k}^2 \tag{4.37}$$

subject to the following constraints:

$$\sum_i P_{i,k} = d_k + \text{Loss}_k - \sum_{g=1} x_{g,k} \tag{4.38}$$

$$P_i^{\min} \leq P_{i,k} \leq P_i^{\max} \tag{4.39}$$

$$-\text{DR}_i \leq P_{i,k+1} - P_{i,k} \leq \text{UR}_i \tag{4.40}$$

$$\text{loss}_k = \sum_i \sum_j P_{k,i} \text{bb}_{i,j} P_{k,j} \tag{4.41}$$

$$\sum_{k=1} \sum_{g=1} z_{g,k} \leq \Omega \tag{4.42}$$

$$\sum_{k=1} x_{g,k} \leq CM_g \tag{4.43}$$

$$\sum_{k=1} \left[ z_{g,k} - \left( q_{1,g} x_{g,k}^2 + q_{2,g} x_{g,k} - q_{2,g} x_{g,k} \theta_g \right) \right] \geq 0 \text{ for } g = 1, \ldots, G \tag{4.44}$$

$$\left.\begin{array}{l} \sum_{k=1} \left[ z_{g,k} - \left( q_{1,g} x_{g,k}^2 + q_{2,g} x_{g,k} - q_{2,g} x_{g,k} \theta_g \right) \right] \geq \\ \sum_{k=1} \left[ z_{g-1,k} - \left( q_{1,g-1} x_{g-1,k}^2 + q_{2,g-1} x_{g-1,k} - q_{2,g-1} x_{g-1,k} \theta_{g-1} \right) \right] \end{array} \right\} \text{for } g = 2, \ldots, G \tag{4.45}$$

The desired variable results to be obtained by the optimization model are $z_{g,k}$, $x_{g,k}$, and $P_{i,k}$.

Equations (4.36) and (4.37) are for generating fuel cost and emission cost, respectively, and both are quadratic functions of the power output of the generators [16]. Equation (4.38) is the supply–demand balance constraint and it ensures that at any time period t, the total power produced by the generators equals the load demand. Constraint (4.39) gives the generation limits and states that the generator limits must not be exceeded. Equation (4.40) is the generator ramp rate constraint and ensures that the ramp rate limits are not exceeded. Equation (4.41) defines the transmission losses constraint. Constraint (4.42) is the utility budget limit and ensures that the total incentive paid is less than the budget. Equation (4.43) is the curtailment constraint and states that at each customer side, the total daily power curtailed must be less than the interruptible power limit. Constraint (4.44) states that the total customer incentive received must be greater or equal to the interruption

**Table 4.11** Parameters for the six generators

| Generators | $a_i$ | $b_i$ | $c_i$ | $e_i$ | $f_i$ | $g_i$ | $P_i^{min}$ | $P_i^{max}$ | $DR_i$ | $UR_i$ |
|---|---|---|---|---|---|---|---|---|---|---|
| 1 | 240 | 7 | 0.007 | 13.8593 | 0.32767 | 0.00419 | 100 | 500 | 120 | 80 |
| 2 | 200 | 10 | 0.0095 | 13.8593 | 0.32767 | 0.00419 | 50 | 200 | 90 | 50 |
| 3 | 220 | 8.5 | 0.009 | 40.2669 | −0.54551 | 0.00683 | 80 | 300 | 100 | 65 |
| 4 | 200 | 11 | 0.009 | 40.2669 | −0.54551 | 0.00683 | 50 | 150 | 90 | 50 |
| 5 | 220 | 10.5 | 0.008 | 42.8955 | −0.51116 | 0.00461 | 50 | 200 | 90 | 50 |
| 6 | 190 | 12 | 0.0075 | 42.8955 | −0.51116 | 0.00461 | 50 | 120 | 90 | 50 |

costs daily. Constraint (4.45) states that the more power is curtailed, the greater the benefit to the customer.

This incentive-based DEED problem will be validated on six generating units with five aggregated customers. Table 4.11 presents generator data for the fuel and emission costs as obtained in [17]. Table 4.12 gives the hourly load demand by the customers. Table 4.13 shows the coefficients for the cost function, customer power limit, and customer type.

**Table 4.12** Hourly load demand

| Hour | Load demand |
|---|---|
| 1 | 955 |
| 2 | 942 |
| 3 | 935 |
| 4 | 930 |
| 5 | 935 |
| 6 | 963 |
| 7 | 989 |
| 8 | 1023 |
| 9 | 1126 |
| 10 | 1150 |
| 11 | 1201 |
| 12 | 1235 |
| 13 | 1190 |
| 14 | 1251 |
| 15 | 1263 |
| 16 | 1250 |
| 17 | 1221 |
| 18 | 1202 |
| 19 | 1159 |
| 20 | 1092 |
| 21 | 1023 |
| 22 | 984 |
| 23 | 975 |
| 24 | 960 |

**Table 4.13** Power produced by generators

| Hour | $P_1$ | $P_2$ | $P_3$ | $P_4$ | $P_5$ | $P_6$ |
|---|---|---|---|---|---|---|
| 1 | 282.56 | 121.4 | 179.94 | 100.98 | 145.23 | 89.29 |
| 2 | 279.47 | 118.87 | 177.75 | 98.79 | 142.48 | 86.43 |
| 3 | 278.38 | 117.98 | 176.99 | 98.02 | 141.52 | 85.43 |
| 4 | 277.46 | 117.22 | 176.34 | 97.37 | 140.7 | 84.58 |
| 5 | 277.58 | 117.32 | 176.42 | 97.45 | 140.8 | 84.68 |
| 6 | 282 | 120.93 | 179.54 | 100.58 | 144.72 | 88.76 |
| 7 | 273.23 | 113.77 | 173.35 | 94.38 | 136.95 | 80.67 |
| 8 | 281.69 | 120.68 | 179.33 | 100.36 | 144.45 | 88.49 |
| 9 | 294.38 | 131.06 | 188.3 | 109.33 | 155.72 | 100.21 |
| 10 | 309.17 | 143.14 | 198.75 | 119.79 | 168.84 | 113.88 |
| 11 | 318.67 | 150.91 | 205.47 | 126.5 | 177.27 | 120 |
| 12 | 331.14 | 161.1 | 214.28 | 135.31 | 188.33 | 120 |
| 13 | 326.31 | 157.15 | 210.86 | 131.9 | 184.04 | 120 |
| 14 | 335.06 | 164.3 | 217.05 | 138.09 | 191.81 | 120 |
| 15 | 344.1 | 171.7 | 223.44 | 144.48 | 199.84 | 120 |
| 16 | 340.62 | 168.85 | 220.98 | 142.01 | 196.74 | 120 |
| 17 | 327.29 | 157.95 | 211.56 | 132.59 | 184.91 | 120 |
| 18 | 323.92 | 155.2 | 209.18 | 130.21 | 181.93 | 120 |
| 19 | 315.81 | 148.57 | 203.44 | 124.48 | 174.73 | 120 |
| 20 | 301.91 | 137.2 | 193.61 | 114.65 | 162.39 | 107.16 |
| 21 | 290.42 | 127.81 | 185.49 | 106.53 | 152.19 | 96.55 |
| 22 | 281.3 | 120.36 | 179.05 | 100.09 | 144.11 | 88.13 |
| 23 | 284.8 | 123.23 | 181.53 | 102.56 | 147.21 | 91.36 |
| 24 | 282.88 | 121.65 | 180.16 | 101.2 | 145.5 | 89.58 |

The optimization model is formulated and solved using the CONPOT solver embedded in AIMMS. The simulation results are presented in Tables 4.13, 4.14 and 4.15 and Figs. 4.6, 4.7 and 4.8.

Table 4.13 presents the optimal power output from the six generators and the graphical view is shown in Fig. 4.6. It is observed that the generators reduce their power output when customers willingly reduce their demands. Tables 4.14 and 4.15 give the optimal power curtailed by customers and the incentive paid to these customers, respectively. The results show that the incentive paid to customers alters the pattern load demand by the customer, which in turn improves the power system security and reliability.

**Table 4.14** Power curtailed by customer

| Hour | $g_1$ | $g_2$ | $g_3$ | $g_4$ | $g_5$ |
|---|---|---|---|---|---|
| 1 | 3.64 | 5.61 | 9.1 | 11.76 | 14.67 |
| 2 | 3.93 | 5.99 | 9.58 | 12.36 | 15.25 |
| 3 | 3.74 | 5.73 | 9.24 | 11.97 | 14.8 |
| 4 | 3.72 | 5.38 | 9.2 | 11.99 | 14.75 |
| 5 | 4.29 | 5.9 | 10.18 | 13.14 | 15.96 |
| 6 | 5.02 | 7.01 | 11.46 | 14.57 | 17.55 |
| 7 | 13.49 | 18 | 26.27 | 31.27 | 35.98 |
| 8 | 12.43 | 17.41 | 24.47 | 29.12 | 33.67 |
| 9 | 17.31 | 23.97 | 33.02 | 38.76 | 44.25 |
| 10 | 11.36 | 16 | 22.55 | 27.1 | 31.22 |
| 11 | 12.22 | 17.11 | 23.95 | 28.8 | 32.87 |
| 12 | 10.25 | 14.51 | 20.46 | 25.05 | 28.51 |
| 13 | 7.14 | 10.32 | 15.09 | 18.8 | 21.86 |
| 14 | 10.27 | 14.52 | 20.54 | 25.01 | 28.65 |
| 15 | 7.3 | 10.52 | 15.42 | 19.04 | 22.37 |
| 16 | 7.4 | 10.66 | 15.66 | 19.19 | 22.75 |
| 17 | 10.4 | 14.44 | 20.95 | 25.12 | 29.36 |
| 18 | 10.15 | 13.29 | 20.14 | 22.98 | 28.26 |
| 19 | 8.52 | 12.08 | 17.59 | 21.18 | 25.11 |
| 20 | 8.71 | 12.39 | 17.98 | 21.45 | 25.6 |
| 21 | 7.18 | 10.4 | 15.23 | 18.94 | 22.18 |
| 22 | 7.92 | 11.43 | 16.44 | 20.59 | 23.66 |
| 23 | 4.73 | 7.11 | 10.91 | 14.12 | 16.82 |
| 24 | 4.07 | 6.21 | 9.78 | 12.75 | 15.44 |

## 4.5 Profit-Based Dynamic Economic Emissions Dispatch

In the profit-based DEED system, the operation and control of the generating units are under the control of the utility companies [18, 19]. The perspective of these utility companies is predominantly to maximize profits, which is expressed as the difference between accumulated revenue and the operational cost of running the generating units. This accumulated revenue is given as the predicted energy prices that utilities gain over the dispatching period [20, 21]. Profit-based DEED is a multi-objective optimization problem that seeks to maximize profit and minimize emissions. The mathematical formulation is expressed as:

$$\text{Min}\left[w_1\left[\sum_{k=1}\sum_{i=1}C_i(P_{k,i})-\sum_{k=1}\sum_{i=1}E_k^{\text{price}}(P_{k,i})\right]+w_2\sum_{k=1}\sum_{i=1}E_i(P_{k,i})\right] \quad (4.46)$$

$$C_i(P_{k,i})=a_i+b_iP_{k,i}+c_iP_{k,i}^2 \quad (4.47)$$

**Table 4.15** Customer incentive

| Hour | $g_1$ | $g_2$ | $g_3$ | $g_4$ | $g_5$ |
|---|---|---|---|---|---|
| 1 | 66.77 | 97.38 | 142.73 | 163.93 | 189.25 |
| 2 | 74.23 | 107.15 | 155.22 | 178.97 | 204.42 |
| 3 | 69.36 | 100.37 | 146.19 | 169.21 | 192.72 |
| 4 | 68.97 | 91.72 | 145.25 | 169.64 | 191.29 |
| 5 | 83.85 | 104.57 | 171.44 | 199.84 | 224.04 |
| 6 | 104.85 | 135.13 | 208.72 | 240.43 | 270.9 |
| 7 | 493.22 | 619.34 | 898.62 | 992.56 | 1138.2 |
| 8 | 430.06 | 584.81 | 789.42 | 867.07 | 996.98 |
| 9 | 755 | 1022.53 | 1369.56 | 1495.06 | 1721.79 |
| 10 | 370.73 | 506.67 | 680.55 | 757.22 | 857.15 |
| 11 | 417.86 | 568.09 | 758.94 | 849.43 | 950.14 |
| 12 | 313.12 | 429.42 | 571.7 | 653.06 | 714.9 |
| 13 | 177.38 | 246 | 334.09 | 383.01 | 420.35 |
| 14 | 314.36 | 430.13 | 575.72 | 651.2 | 721.88 |
| 15 | 183.3 | 253.79 | 346.86 | 391.88 | 439.96 |
| 16 | 187.43 | 259.17 | 356.34 | 397.59 | 455.04 |
| 17 | 320.93 | 426.3 | 596.22 | 656.51 | 757.93 |
| 18 | 308.29 | 371.06 | 555.8 | 555.61 | 702.35 |
| 19 | 233.26 | 317.23 | 436.7 | 477.43 | 554.32 |
| 20 | 241.4 | 330.81 | 454.14 | 488.53 | 576.21 |
| 21 | 178.82 | 249.18 | 339.32 | 388.05 | 432.79 |
| 22 | 207.89 | 289.72 | 387.94 | 452.73 | 492.38 |
| 23 | 96.45 | 138.15 | 192.33 | 227.32 | 248.68 |
| 24 | 77.88 | 112.79 | 160.6 | 189.18 | 209.53 |

$$E_i(P_{k,i}) = e_i + f_i P_{k,i} + g_i P_{k,i}^2 \tag{4.48}$$

$$w_1 + w_2 = 1 \tag{4.49}$$

subject to the following constraints:

$$\sum_i P_{k,i} \leq d_k + \text{loss}_k \tag{4.50}$$

$$\text{loss}_k = \sum_i \sum_j P_{k,i} \text{bb}_{i,j} P_{k,j} \tag{4.51}$$

$$P_i^{\min} \leq P_{k,i} \leq P_i^{\max} \tag{4.52}$$

$$P_{k,i} - P_{(k+1),i} \leq \mathrm{DR}_i \tag{4.53}$$

$$P_{k,i} - P_{(k+1),i} \leq \mathrm{UR}_i. \tag{4.54}$$

The desired variable result to be obtained by the optimization model is $P_{k,i}$.

The quadratic cost function for the fuel and emissions is presented in Eqs. (4.47) and (4.48). Equation (4.49) is the weighting factor for the fuel cost and emission functions. Constraint (4.50) is the supply–demand balance equation and states that the power produced in the generating units can be less than the total loads. Equation (4.51) is known as the B-coefficient method and describes the best approach to calculate losses. Constraint (4.52) describes the generation limits and ensures that these limits do not exceed the stipulated limit. Constraints (4.53) and (4.54) are the generator ramp rates.

The optimization model is formulated and solved using the CONPOT solver embedded in AIMMS. The AIMMS code for solving an example of a profit-based DEED problem is shown below.

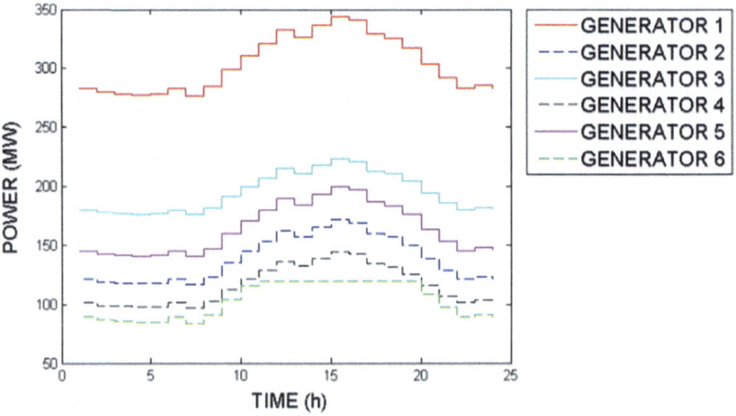


**Fig. 4.6** Output power generated by the generators

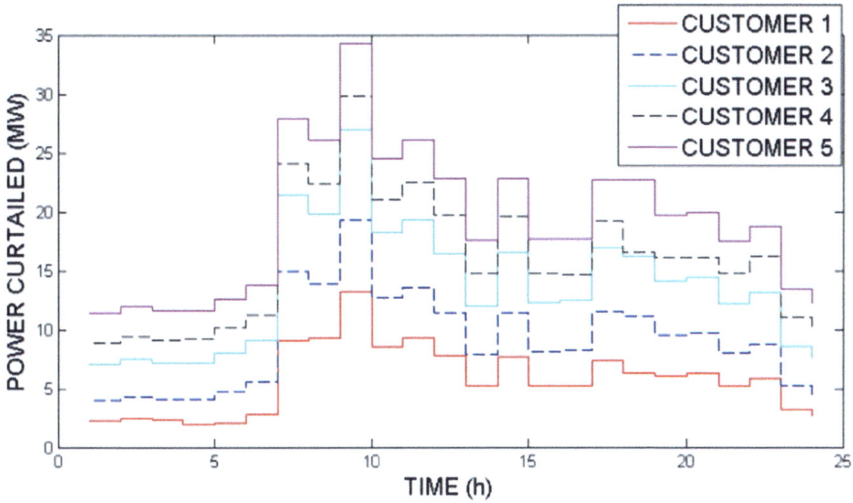


**Fig. 4.7** Power curtailed by the customers

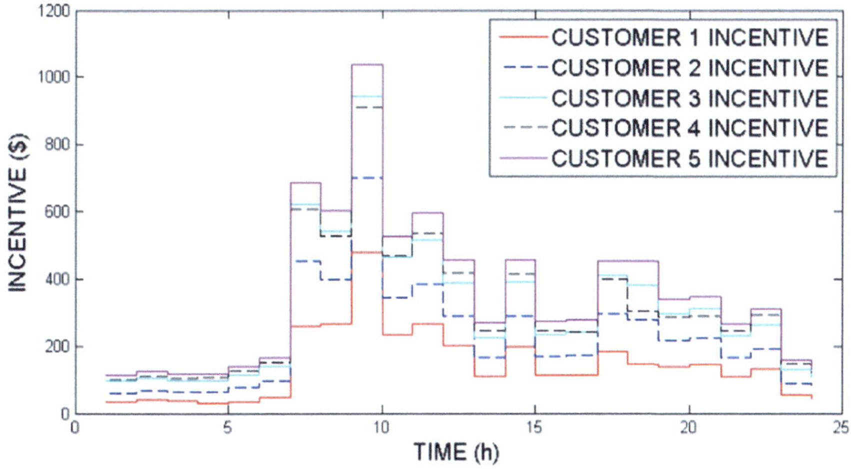


**Fig. 4.8** Incentive paid to the customers

The AIMMS code for solving the profit-based DEED is given below:

```
Model Main_Profit_based_DEED {
  Set Generators {
    Index: i, j;
  }
  Set Time {
    Index: k, kk, kkk;
  }
  Parameter pmin {
    IndexDomain: i;
  }
  Parameter pmax {
    IndexDomain: i;
  }
  Parameter a {
    IndexDomain: i;
  }
  Parameter e {
    IndexDomain: i;
  }
  Parameter f {
    IndexDomain: i;
  }
  Parameter g {
    IndexDomain: i;
  }
  Parameter y {
    Definition: 0;
  }
  Parameter z {
    Definition: 1-y;
  }
  Parameter b {
    IndexDomain: i;
  }
  Parameter c {
    IndexDomain: i;
  }
  Parameter ur {
    IndexDomain: i;
  }
  Parameter dr {
    IndexDomain: i;
  }
  Parameter d {
    IndexDomain: k;
  }
  Parameter bb {
    IndexDomain: (i,j);
  }
  Parameter loss {
    IndexDomain: k;
    Definition: sum[i,sum[j, p(k,i)*bb(i,j)*p(k,j)]];
  }
  Parameter EP {
    IndexDomain: (k);
  }
  Variable p {
    IndexDomain: (k,i);
    Range: nonnegative;
  }
  Variable Total {
    Range: free;
    Definition: (y*(sum[ (k,i), a(i)+(b(i)*p(k,i))+(c(i)*p(k,i)*p(k,i)) ]-sum[ (k,i), EP(k)*p(k,i) ]))+(z*sum[ (k,i),
e(i)+(f(i)*p(k,i))+(g(i)*p(k,i)*p(k,i)) ]);
  }
  Parameter tloss {
    Definition: sum[k,loss(k)];
```

```
}
Parameter costt {
   Definition: (sum[ (k,i), (a(i)+(b(i)*p(k,i))+(c(i)*p(k,i)*p(k,i)))]);
}
Parameter emissions {
   Definition: (sum[ (k,i), e(i)+(f(i)*p(k,i))+(g(i)*p(k,i)*p(k,i))]);
}
Parameter totgenpower {
   Definition: sum[ (i,k),p(k,i)];
}
Constraint sup {
   IndexDomain: k;
   Definition: sum[i, p(k,i)]<=d(k)+loss(k);
}
MathematicalProgram LeastEconDis {
   Objective: Total;
   Direction: minimize;
   Constraints: AllConstraints;
   Variables: AllVariables;
   Type: Automatic;
}
Constraint sup1 {
   IndexDomain: (k,i);
   Definition: p(k,i)<=pmax(i);
}
Constraint sup2 {
   IndexDomain: (k,i);
   Definition: p(k,i)>=pmin(i);
}
Constraint sup3 {
   IndexDomain: (kk,i) | 1<= ord(kk) <=23;
   Definition: p(kk,i)-p(kk+1,i)<=dr(i);
}
Constraint sup4 {
   IndexDomain: (kk,i) | 1<= ord(kk) <=23;
   Definition: p(kk+1,i)-p(kk,i)<=ur(i);
}
Procedure MainInitialization;
Procedure MainExecution {
   Body: {
      solve LeastEconDis;
   }
}
```

Table 4.16 provides the data for fuel cost and emission coefficients for the six generators, as obtained in [17]. Figures 4.9 and 4.10 describe the hourly load profile and the forecast energy prices for the day. Table 4.17 provides the numerical results for the profit-based DEED model. The results show the optimal power generated by all six generators.

**Table 4.16** Generator data for profit-based DEED model

| Generator | $a_i$ | $b_i$ | $c_i$ | $e_i$ | $f_i$ | $g_i$ | $P_i^{min}$ | $P_i^{max}$ | $R_i^{down}$ | $R_i^{up}$ |
|---|---|---|---|---|---|---|---|---|---|---|
| 1 | 240 | 7 | 0.007 | 13.8593 | 0.32767 | 0.00419 | 100 | 500 | 120 | 80 |
| 2 | 200 | 10 | 0.0095 | 13.8593 | 0.32767 | 0.00419 | 50 | 200 | 90 | 50 |
| 3 | 220 | 8.5 | 0.009 | 40.2669 | −0.54551 | 0.00683 | 80 | 300 | 100 | 65 |
| 4 | 200 | 11 | 0.009 | 40.2669 | −0.54551 | 0.00683 | 50 | 150 | 90 | 50 |
| 5 | 220 | 10.5 | 0.008 | 42.8955 | −0.51116 | 0.00461 | 50 | 200 | 90 | 50 |
| 6 | 190 | 12 | 0.0075 | 42.8955 | −0.51116 | 0.00461 | 50 | 120 | 90 | 50 |

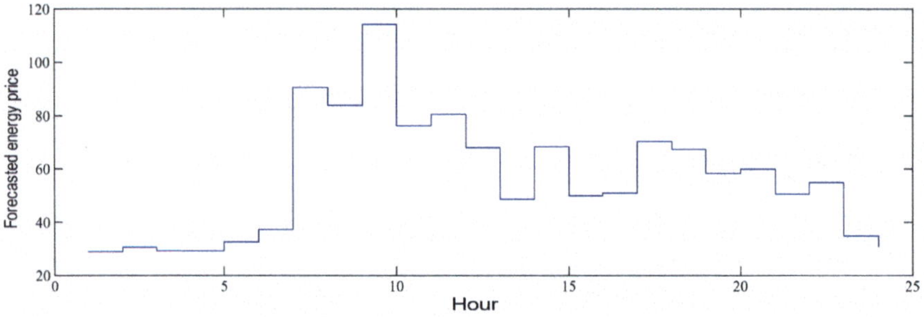


**Fig. 4.9** The forecast energy price for the profit-based DEED model

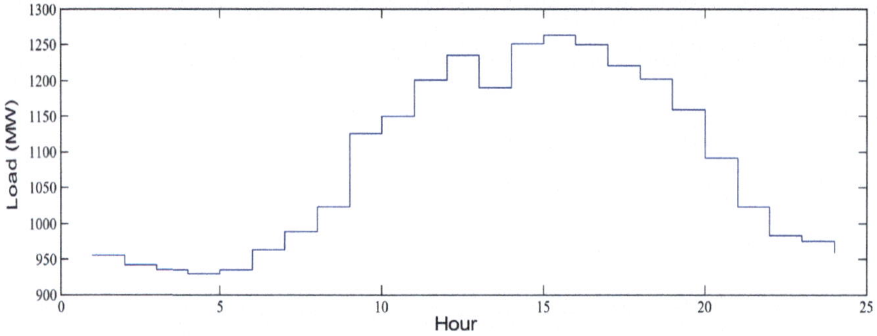


**Fig. 4.10** Total hourly load for the profit-based DEED model

## 4.6 Summary

In this chapter, different formulations of DEED problems with inclusion of DR programs were investigated. The objective of the models is to determine the optimal power produced by the generators while minimizing the fuel cost and emission cost, as well as maximizing the incentives paid to the customers. The results obtained show that the inclusion of DR programs in DEED help to reduce the total load requested by customers.

**Table 4.17** Optimal generated power for the profit-based DED model

| Time (hr) | Gen 1 | Gen 2 | Gen 3 | Gen 4 | Gen 5 | Gen 6 |
|---|---|---|---|---|---|---|
| 1 | 100 | 50 | 80 | 50 | 55.44 | 55.44 |
| 2 | 100 | 50 | 80 | 50 | 55.44 | 55.44 |
| 3 | 100 | 50 | 80 | 50 | 55.44 | 55.44 |
| 4 | 100 | 50 | 80 | 50 | 55.44 | 55.44 |
| 5 | 100 | 50 | 80 | 50 | 55.44 | 55.44 |
| 6 | 100 | 50 | 80 | 50 | 55.44 | 55.44 |
| 7 | 100 | 50 | 80 | 50 | 55.44 | 55.44 |
| 8 | 100 | 50 | 80 | 50 | 55.44 | 55.44 |
| 9 | 100 | 50 | 80 | 50 | 55.44 | 55.44 |
| 10 | 100 | 50 | 80 | 50 | 55.44 | 55.44 |
| 11 | 100 | 50 | 80 | 50 | 55.44 | 55.44 |
| 12 | 100 | 50 | 80 | 50 | 55.44 | 55.44 |
| 13 | 100 | 50 | 80 | 50 | 55.44 | 55.44 |
| 14 | 100 | 50 | 80 | 50 | 55.44 | 55.44 |
| 15 | 100 | 50 | 80 | 50 | 55.44 | 55.44 |
| 16 | 100 | 50 | 80 | 50 | 55.44 | 55.44 |
| 17 | 100 | 50 | 80 | 50 | 55.44 | 55.44 |
| 18 | 100 | 50 | 80 | 50 | 55.44 | 55.44 |
| 19 | 100 | 50 | 80 | 50 | 55.44 | 55.44 |
| 20 | 100 | 50 | 80 | 50 | 55.44 | 55.44 |
| 21 | 100 | 50 | 80 | 50 | 55.44 | 55.44 |
| 22 | 100 | 50 | 80 | 50 | 55.44 | 55.44 |
| 23 | 100 | 50 | 80 | 50 | 55.44 | 55.44 |
| 24 | 100 | 50 | 80 | 50 | 55.44 | 55.44 |

## 4.7 Nomenclature

The nomenclature used in this chapter is given below:

**Set**

$g$ Customer
i Generator
$k$ Time

**Parameters**

$bb_{i,j}$ Loss coefficient square matrix
$B_g^{\min}$ Lower quantities of bid for customer g
$B_g^{\max}$ Upper quantities of bid for customer $g$

| | |
|---|---|
| $C_i$ | Fuel cost of generator *i* |
| $d_k$ | Load |
| $E_i$ | Emissions cost generator *i* |
| $E^{\max}$ | Maximum permissible emissions for the generators |
| $E_k^{\text{price}}$ | Predicted price of energy at time *k* |
| $H_i$ | Emission function of generator *i* |
| $P_i^{\min}$ | Lower capacity of generator *i* |
| $P_i^{\max}$ | Upper capacity of generator i |
| $\text{loss}_k$ | System losses |
| $\text{DR}_i$ | Ramp-down rate of generator *i* |
| $\text{UR}_i$ | Ramp-up rate of generator *i* |
| $\delta_g$ | Bid function for customer *g* |
| $\lambda_{g,k}$ | Power interruptible value for different customer g at time k |
| $\Omega$ | Budget for the utility |
| $q_1$ and $q_2$ | Coefficients of cost function for customer *g* |
| $\theta_g$ | Type of customer |
| $w_1$ and $w_2$ | Weighting factor |
| $a_i$, $b_i$ and $c_i$ | Coefficients of fuel cost of generator *i* |
| $e_i$, $f_i$ and $g_i$ | Coefficients of emission cost of generator *i* |

### Variables

| | |
|---|---|
| $B_{g,k}$ | Quantity of bid for customer g at time *k* |
| $P_{i,k}$ | Power produced by generator i at time *k* |
| $x_{g,k}$ | Quantity of curtailed power by customer g at time *k* |
| $z_{g,k}$ | Amount of incentive paid to customer g at time *k* |

## References


1. Nwulu NI, Xia X (2015) Multi-objective dynamic economic emission dispatch of electric power generation integrated with game theory based demand response programs. Energy Convers Manag 89:963–974
2. Lokeshgupta B, Sivasubramani S (2018) "Electrical power and energy systems multi-objective dynamic economic and emission dispatch with demand side management," Electr Power Energy Syst 97(November 2017):334–343
3. Narimani H et al (2018) A multi-objective framework for multi-area economic emission dispatch. Energy 154:126–142
4. Gbadamosi SL, Nwulu NI (2020) "Optimal power dispatch and reliability analysis of hybrid CHP-PV-wind systems in farming applications." Sustainability 12
5. Gbadamosi SL, Nwulu NI (2019) "A dynamic optimal expansion planning model integrating renewable energy sources." In: 11th International conference on applied energy pp 150–155
6. Gbadamosi SL, Nwulu NI (2020) Optimal planning of renewable energy systems for power loss reduction in transmission expansion planning. J Eng Des Technol 18(5):1209–1222

7. Parvez F, Vasant P, Kallimani V, Watada J (2018) "A holistic review on optimization strategies for combined economic emission dispatch problem," Renew Sustain Energy Rev 81 (March 2017):3006–3020
8. Damisa U, Nwulu NI, Sun Y (2018) Microgrid energy and reserve management incorporating prosumer behind-the-meter resources. IET Renew Power Gener 12(8):910–919
9. S. L. Gbadamosi and N. I. Nwulu, "A multi-period composite generation and transmission expansion planning model incorporating renewable energy sources and demand response," Sustain. Energy Technol. Assessments, vol. 39, no. October 2019, p. 100726, 2020
10. Gbadamosi SL, Nwulu NI, Sun Y (2018) Multi-objective optimisation for composite generation and transmission expansion planning considering offshore wind power and feed-in tariffs. IET Renew Power Gener 12(14):1687–1697
11. Gbadamosi SL, Nwulu NI, Sun Y (2018) Harmonic and power loss minimization in power systems incorporating renewable energy sources and locational marginal pricing. J Renew Sustain Energy 10:1–24
12. Gbadamosi SL, Nwulu NI, Sun Y (2018) "Optimal scheduling in offshore wind farms considering harmonic losses and locational marginal pricing." 2018 IEEE PES-IAS 5th Power Africa Conf, pp 658–663
13. Basu M (2014) Fuel constrained economic emission dispatch using nondominated sorting genetic algorithm-II. Energy 78:649–664
14. Nwulu NI (2017) "Emission constrained bid based dynamic economic dispatch using quadratic programming," In 2017 International conference on energy, communication, data analytics and soft computing (ICECDS), pp. 213–216
15. Gbadamosi SL, Nwulu NI (2020) "Reliability assessment of composite generation and transmission expansion planning incorporating renewable energy sources." J Renew Sustain Energy 026301(12)
16. Nwulu NI, Xia X (2015) Implementing a model predictive control strategy on the dynamic economic emission dispatch problem with game theory based demand response programs. Energy 91:404–419
17. Fan M, Zhang Z, Wang C (2019) "Optimization of annual generator maintenance scheduling." Academic Press, pp 49–80
18. Ekpenyong UE, Zhang J, Xia X (2012) "An improved robust model for generator maintenance scheduling." Electr Power Syst Res 92
19. Reihani E, Sarikhani A, Davodi M, Davodi M (2012) Reliability based generator maintenance scheduling using hybrid evolutionary approach. Int J Electr Power Energy Syst 42(1):434–439
20. Elaiw AM, Xia X, Shehata AM (2013) "Combined heat and power dynamic economic dispatch with emission limitations using hybrid DE-SQP method." Hindawi Publ Corp, pp 13–15
21. Nwulu NI, Xia X (2017) Optimal dispatch for a microgrid incorporating renewables and demand response. Renew Energy 101:16–28

# Chapter 5
# Generator Maintenance Scheduling

## 5.1 Introduction

This chapter discusses the fundamental role of generator maintenance in providing adequate and reliable electric energy service delivery to consumers. GMS helps in reducing the cost of generation, prolongs the generator's working life and improves the system reliability [1]. The GMS model has to generate a specific time horizon for compiling a preventive maintenance schedule for a certain set of units within a specific period with the aim of optimizing a set of objectives without violating any constraints [2]. An appropriate maintenance schedule should minimize the operating and maintenance cost of the power-generating system or improve the reliability of the system. GMS problems can be categorized into a reliability criterion and an economic cost criterion. Here, we provide solutions for the operating cost and reliability criterion problems for models in AIMMS.

## 5.2 Reliability Criterion Generator Maintenance Scheduling

The reliability criterion aims to ensure adequate balance between the power supply and demand at all times, regardless of load variations. To achieve this, the utility generally provides a spinning reserve by generating more power than demanded, which improves system reliability at the expense of operational cost [3]. The reliability criterion GMS (RC-GMS) problem is solved by minimizing the sum of square of the reserve over the whole operational planning period [2]. During scheduled maintenance outage, fear of insufficient supply from the generators may be increased [4, 5]. Therefore, the RC-GMS problem is formulated to minimize the sum squares of reserve of the generating units.



N. Nwulu and S. L. Gbadamosi, *Optimal Operation and Control of Power Systems Using an Algebraic Modelling Language*, Green Energy and Technology,
https://doi.org/10.1007/978-3-030-00395-1_5

### 5.2.1 Formulation of RC-GMS Model

Here, we consider the reliability the objective and the aim of this objective function is to minimize the sum of squares of reserve of the generating units while satisfying maintenance window constraints:

$$\mathrm{Min}\left[\sum_{t=1}^{T}\left(\sum_{i=1}^{I} G_{i,t}^{\max} - \sum_{i=1}^{I}\sum_{t=1}^{T} G_{i,t}x_{i,t} - D_t\right)^2\right] \tag{5.1}$$

subject to the maintenance constraints given below:

$$\sum_{t \in T} x_{i,t} = d_i \tag{5.2}$$

$$\sum_{t} y_{i,t} = 1 \tag{5.3}$$

$$x_{i,t} - x_{i,t-1} \leq y_{i,t} \tag{5.4}$$

$$\sum_{i \in I} G_{i,t} - \sum_{i \in I}\sum_{t \in T} G_{i,t}x_{i,t} \geq D_t + R_t \tag{5.5}$$

$$G_i^{\min} \leq G_{i,t} \leq G_i^{\max} \tag{5.6}$$

$$\sum_{i \in I} q_{i,t}x_{i,t} \leq Q_t \tag{5.7}$$

The expected desired variable results for the optimization model are: $G_{i,t}$, $x_{i,t}$, $R_t$, and $y_{i,w}$.

The maintenance constraints are briefly described below:

Constraint (5.2) states that maintenance will be for the periods of $d_i$ for all generators i. Equation (5.3) is the generator start-up maintenance constraint and ensures that the maintenance constraint operates at the optimized level. Equation (5.4) stands for the uninterrupted maintenance constraint. Equation (5.5) is the supply balance constraint and ensures that the generated power obtainable from the generating units minus power outage due to maintenance interruption is greater than the demand at all time. Equation (5.6) is the generator limit and ensures that generators remain within their limit boundary [6, 7]. Equation (5.7) is the crew constraint.

**Table 5.1** Data for the 21-generator units test system

| Generating unit | Max Gen (MW) | Min Gen (MW) | Maintenance duration (weeks) | Allowed period |
|---|---|---|---|---|
| 1 | 555 | 462 | 7 | 1–26 |
| 2 | 180 | 150 | 2 | 1–26 |
| 3 | 180 | 150 | 1 | 1–26 |
| 4 | 640 | 533 | 3 | 1–26 |
| 5 | 640 | 533 | 3 | 1–26 |
| 6 | 276 | 230 | 10 | 1–26 |
| 7 | 140 | 117 | 4 | 1–26 |
| 8 | 90 | 75 | 1 | 1–26 |
| 9 | 76 | 63 | 2 | 1–26 |
| 10 | 94 | 78 | 4 | 1–26 |
| 11 | 39 | 32 | 2 | 1–26 |
| 12 | 188 | 152 | 2 | 1–26 |
| 13 | 52 | 43 | 3 | 1–26 |
| 14 | 555 | 462 | 5 | 27–52 |
| 15 | 640 | 533 | 5 | 27–52 |
| 16 | 555 | 462 | 6 | 27–52 |
| 17 | 76 | 63 | 3 | 27–52 |
| 18 | 58 | 48 | 1 | 27–52 |
| 19 | 48 | 40 | 2 | 27–52 |
| 20 | 137 | 114 | 1 | 27–52 |
| 21 | 469 | 392 | 4 | 27–52 |

The developed model is validated on 21 generating units over a target period of 52 weeks and the data can be seen in Table 5.1 [2]. It is assumed that maintenance must be performed on all the generators and ideally each generator should undergo maintenance within the 26-week period [8, 9].

The AIMMS code for solving the example for the RC-GMS model is given below:

```
Model Main_GMS {
  Set Generators {
    Index: i;
    Comment: "Set of all generating units";
  }
  Set TimePeriods {
    Index: t, tt;
    Comment: "Set of Time Periods (Week number)";
  }
  Parameter Gmax {
    IndexDomain: i;
    Comment: "Maximum output power for each generator unit";
  }
  Parameter Gmin {
    IndexDomain: i;
    Comment: "Minimum output power for each generator unit";
  }
  Parameter MaintenanceDuration {
    IndexDomain: i;
    Comment: "Number of time periods required for maintenance of each generator";
  }
  Parameter Demand {
    Comment: "Electricity Demand";
  }
  Parameter DR {
    Comment: "Generator down rate";
  }
  Parameter UR {
    Comment: "Generator Up Rate";
  }
  Parameter AllowedWindow {
    IndexDomain: (i,t);
    Range: binary;
    Comment: "The periods within which maintenance of each generator can take place";
  }
  Parameter ManpowerRequired {
    IndexDomain: (i,t);
    Comment: "The number of personnel required for each stage of maintenance during the maintenance period";
  }
```

```
Variable OnMaintenance {
  IndexDomain: (i,t)|AllowedWindow(i,t);
  Range: binary;
  Comment: "Shows the time periods when a generator will be on maintenance";
}
Variable Gout {
  IndexDomain: (i,t);
  Range: nonnegative;
  Comment: "Generator Output of each generator at each time period";
}
Variable SSRMin {
  Range: free;
  Definition: {
    sum[t,(Reserve(t))^2];
  }
}
Variable Reserve {
  IndexDomain: t;
  Range: free;
  Definition: {
    sum[i,Gmax(i)-(OnMaintenance(i,t)*Gmax(i))]-Demand;
  }
  Comment: "Reserve Margin at each time period";
}
Variable TotalGeneration {
  IndexDomain: t;
  Range: free;
  Definition: {
    sum[i,Gout(i,t)];
  }
  Comment: "Total generated output during each time period";
}
Variable CapacityOnMaintenance {
  IndexDomain: t;
  Range: free;
```

```
    Definition: {
      sum[i,OnMaintenance(i,t)*Gmax(i)];
    }
    Comment: "Total capacity of generators that are on maintenance";
  }
  Variable MaintenanceStart {
    IndexDomain: (i,t);
    Range: binary;
    Comment: "Time periods in which maintenance of a generator starts";
  }
  Variable ManpowerUsed {
    IndexDomain: (i,t);
    Range: integer;
    Definition: {
      sum( tt | val(t) - MaintenanceDuration(i) + 1 <= val(tt) and tt <= t, MaintenanceStart(i,tt) * ManpowerRequired(i,t-
val(tt)+1) );
    }
    Comment: "Number of personnel used at each stage of the maintenance";
  }
  Variable TotalManpower {
    IndexDomain: t;
    Range: nonnegative;
    Definition: {
      sum(i,ManpowerUsed(i,t));
    }
    Comment: "Total personnel used for maintenance at each time period";
  }
  Variable TMW {
    Range: free;
    Definition: {
      sum(t,TotalManpower(t));
    }
    Comment: "Total Man Weeks";
  }
  Constraint MaintDurationConst {
    IndexDomain: i;
    Definition: {
      sum[t,OnMaintenance(i,t)]=MaintenanceDuration(i);
    }
    Comment: "Maintenance Duration Constraint";
```

```
}
Constraint ManpowerConstraint {
  IndexDomain: t;
  Definition: {
    sum(i,ManpowerUsed(i,t))<=40;
  }
  Comment: "Total Man-Week constraint";
}
Constraint Uninterrupt {
  IndexDomain: (i,t);
  Definition: {
    OnMaintenance(i,t)-OnMaintenance(i,t-1) <= MaintenanceStart(i,t);
  }
  Comment: "ensure maintenance is uninterrupted";
}
Constraint UninterruptA {
  IndexDomain: i;
  Definition: {
    sum(t,MaintenanceStart(i,t))<=1;
  }
    Comment: "ensure maintenance is uninterrupted";
  }
  Constraint LoadConstraint {
    IndexDomain: t;
    Definition: {
      sum[i,Gout(i,t)]>=Demand;
    }
    Comment: "Ensure total output per time period is greater than demand";
  }
  Constraint GoutMin {
    IndexDomain: (i,t);
    Definition: {
      Gout(i,t)>=Gmin(i)*(1-OnMaintenance(i,t));
    }
```

```
    Comment: "Minimimum generator output constraint";
}
Constraint GoutMax {
    IndexDomain: (i,t);
    Definition: {
        Gout(i,t)<=Gmax(i)*(1-OnMaintenance(i,t));
    }
    Comment: "Maximum generator output constraint";
}
MathematicalProgram MinSSR {
    Objective: SSRMin;
    Direction: minimize;
    Constraints: AllConstraints;
    Variables: AllVariables;
    Type: Automatic;
}
Procedure MainInitialization {
    Comment: "Add initialization statements here that do NOT require any library being initialized already."
}
Procedure PostMainInitialization {
    Comment: {
        "Add initialization statements here that require that the libraries are already initialized properly,
        or add statements that require the Data Management module to be initialized."
    }
 }
 Procedure MainExecution {
    Body: {
        solve MinSSR;

        if (MinSSR.ProgramStatus <> 'Optimal') then
            empty MinSSR;
        endif;
    }
 }
```

The result obtained from the simulation gives the optimal square of the reserve as $13.282 \times 10^6$ with total available man-weeks as 695. Figure 5.1 shows the graph for crew availability for GMS and the total available generation is given in Fig. 5.2, which is within the limit of the generating capacity. Figure 5.3 illustrates the reserve margin and maintenance capacity for GMS.

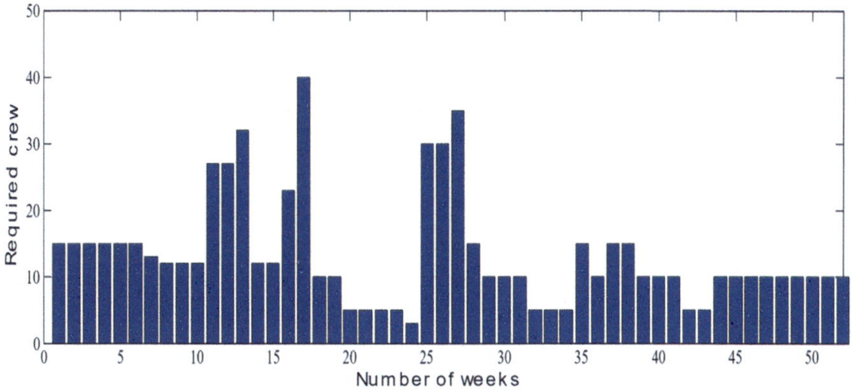


**Fig. 5.1** Crew availability for maintenance

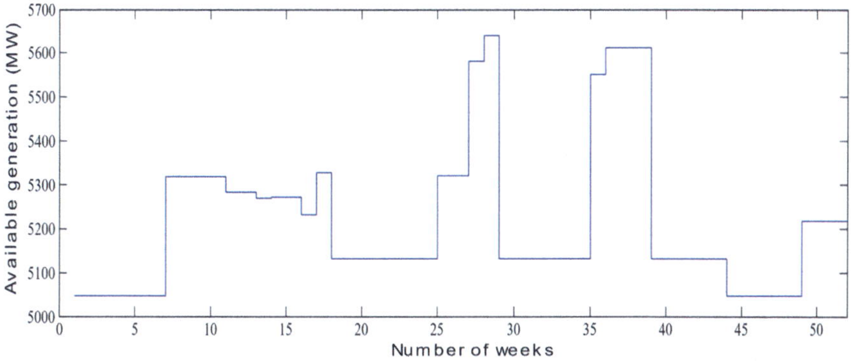


**Fig. 5.2** Total available generation

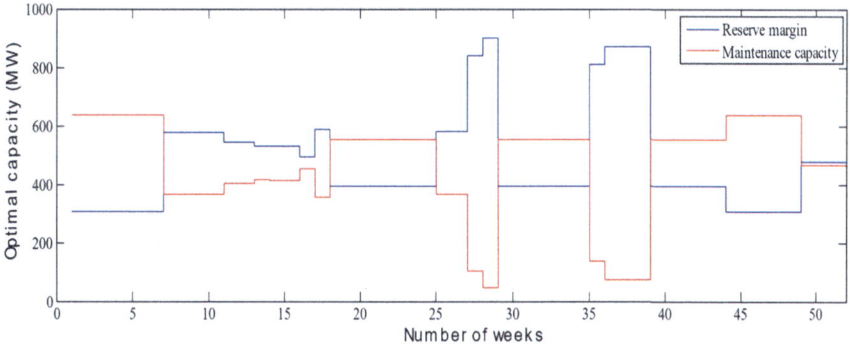


**Fig. 5.3** Reserve margin and maintenance capacity for GMS

## 5.3 Summary

This chapter provides analysis of RC-GMS, with the aim of minimizing the sum squares of reserve of the generating units without violating maintenance window constraints. The model example developed was solved using AIMMS and the simulated results will ensure high system reliability with adequate generation availability and personnel crew.

## 5.4 Nomenclature

The nomenclature used in this chapter is given below:

**Set**

$i$ generator
$t$ time

**Parameter**

$d_i$ maintenance duration for generator $i$
$D_t$ load demand at time $t$
$Q_t$ maximum crew present at time $t$
$G_i^{\min}$ minimum capacity of generator $i$
$G_i^{\max}$ maximum capacity of generator $i$

**Variable**

$G_{i,t}$ power produced by generator $i$ at time $t$
$q_{i,t}$ available crew for generator $i$ maintenance at time $t$
$R_t$ reserve margin at time $t$
$x_{i,t}$ binary variable equals 1 when generator $i$ at time $t$ is on maintenance
$y_{i,w}$ binary variable equals 1 if maintenance of generator $i$ starts at time $t$

## References

1. Nwulu NI, Xia X (2015) Multi-objective dynamic economic emission dispatch of electric power generation integrated with game theory based demand response programs. Energy Convers Manag 89:963–974
2. Zhang N, Hu Z, Springer C, Li Y, Shen B (2016) A bi-level integrated generation-transmission planning model incorporating the impacts of demand response by operation simulation. Energy Convers Manag 123:84–94
3. Rouhani A, Hosseini SH, Raoofat M (2014) Composite generation and transmission expansion planning considering distributed generation. Int J Electr Power Energy Syst 62:792–805

4. Hemmati R, Hooshmand RA, Khodabakhshian A (2013) Reliability constrained generation expansion planning with consideration of wind farms uncertainties in deregulated electricity market. Energy Convers Manag 76:517–526
5. Gbadamosi SL, Nwulu NI, Sun Y (2018) Multi-objective optimisation for composite generation and transmission expansion planning considering offshore wind power and feed-in tariffs. IET Renew Power Gener 12(14):1687–1697
6. Zhang J, Wei Y, Li D, Tan Z, Zhou J (2018) Short term electricity load forecasting using a hybrid model. Energy 158:774–781
7. Gbadamosi SL, Nwulu NI (2020) Reliability assessment of composite generation and transmission expansion planning incorporating renewable energy sources. J Renew Sustain Energy 026301(12)
8. Gbadamosi SL, Nwulu NI, Sun Y (2018) Harmonic and power loss minimization in power systems incorporating renewable energy sources and locational marginal pricing. J Renew Sustain Energy 10:1–24
9. Gbadamosi SL, Nwulu NI, Sun Y (2018) Optimal scheduling in offshore wind farms considering harmonic losses and locational marginal pricing. In: 2018 IEEE PES-IAS 5th power Africa conference, pp 658–663

# Chapter 6
# Combined Heat and Power Dispatch

## 6.1 Introduction

This chapter focusses on the combined heat and power (CHP) DEED (CHPDEED) problem formulation. This problem is a complicated nonlinear mathematical formulation with multiple, conflicting objective functions. The aim of this mathematical problem is to obtain the optimal quantities of heat and power output for the committed generating units, which include power-only and heat-only units. Furthermore, demand response programs such as incentive-based and profit-based DR are also considered with CHPDEED problem formulations. Here, the different models of CHPDEED formulations are solved using AIMMS.

## 6.2 Combined Heat and Power Dynamic Economic Dispatch

CHP-generating units, also known as co-generation units, produce electric power and heat simultaneously [1]. The generation of both power and heat at the output against only power generation significantly increases the efficiency of CHP units. Therefore, CHP units have higher efficiency than conventional generation units. The economic dispatch problem is one of the most important issues in the operation and control of power systems. This problem is defined as determining the optimal power plant generation following the set of different constraints, while keeping the operational cost minimized.



N. Nwulu and S. L. Gbadamosi, *Optimal Operation and Control of Power Systems Using an Algebraic Modelling Language*, Green Energy and Technology,
https://doi.org/10.1007/978-3-030-00395-1_6

The main objective of the CHP dynamic economic dispatch (CHPDED) problem is minimizing the fuel cost. This modeling problem specifies the optimal output power of the CHP-generating units over a daily period minimizing the total operational cost satisfying the unit and system constraints [2, 3].

The total fuel cost (for thermal, CHP and heat units) is given by:

$$\mathrm{Min}C(P,H)=\mathrm{Min}\left[\sum_{k=1}^{k}\sum_{j=1}^{J}C_j^T\left(P_{j,k}^T\right)+\sum_{k=1}^{K}\sum_{g=1}^{G}C_g^{\mathrm{CHP}}\left(P_{g,k}^{\mathrm{CHP}},H_{g,k}^{\mathrm{CHP}}\right)+\sum_{k=1}^{K}\sum_{i=1}^{I}C_i^H\left(H_{i,k}^H\right)\right] \tag{6.1}$$

subject to the following constraints:

$$\sum_{j=1}^{J}P_{j,k}^T+\sum_{g=1}^{G}P_{g,k}^{\mathrm{CHP}}=D_k+\mathrm{Loss}_k \tag{6.2}$$

$$\sum_{g=1}^{G}H_{g,k}^{\mathrm{CHP}}+\sum_{i=1}^{I}H_{i,k}^H=\mathrm{HD}_k+\mathrm{Loss}_k \tag{6.3}$$

$$P_{j,\min}^T\le P_{j,k}^T\le P_{j,\max}^T \tag{6.4}$$

$$P_{g,\min}^{\mathrm{CHP}}\left(H_{g,k}^{\mathrm{CHP}}\right)\le P_{g,k}^{\mathrm{CHP}}\le P_{g,\max}^{\mathrm{CHP}}\left(H_{g,k}^{\mathrm{CHP}}\right) \tag{6.5}$$

$$H_{g,\min}^{\mathrm{CHP}}\left(P_{g,k}^{\mathrm{CHP}}\right)\le H_{g,k}^{\mathrm{CHP}}\le H_{g,\max}^{\mathrm{CHP}}\left(P_{g,k}^{\mathrm{CHP}}\right) \tag{6.6}$$

$$H_{g,\min}^H\le H_{g,k}^H\le H_{g,\max}^H \tag{6.7}$$

$$-\mathrm{DR}_j^T\le P_{j,k+1}^T-P_{j,k}^T\le \mathrm{UR}_j^T \tag{6.8}$$

$$-\mathrm{DR}_g^{\mathrm{CHP}}\le P_{g,t+1}^{\mathrm{CHP}}-P_{g,t}^{\mathrm{CHP}}\le \mathrm{UR}_g^{\mathrm{CHP}} \tag{6.9}$$

$$\mathrm{Loss}_k=\sum_{j=1}^{J}\sum_{z=1}^{Z}P_{j,k}B_{j,z}P_{z,k} \tag{6.10}$$

The desired variable result to be obtained by the optimization model is $P_{j,k}^T$, $P_{g,k}^{\mathrm{CHP}}$, $H_{g,k}^{\mathrm{CHP}}$, $H_{i,k}^H$ and $\mathrm{Loss}_k$.

Equation (6.2) is a power balance constraint, which compels the total output power from both thermal and CHP-generating units to satisfy the load demand and line losses. Constraint (6.3) is a heat balance constraint and it states that the heat output from both heat and CHP units must meet the heat demand. Constraint (6.4) is the generation limits for thermal-generating units and it states that output power from the thermal units must not exceed the allowed limits. The next constraint (6.5) limits the power produced by CHP units within the allowable limits. Constraint (6.6) limits the heat produced by CHP units within the allowable limits. Constraint (6.7) ensures that the heat produced by heat-only units is within the allowable limits. Constraints (6.8) and (6.9) are the generator ramp rate limits for the thermal and CHP-generating units. Equation (6.10) is the transmission line losses constraint and the line losses are determined by the B-coefficient method [4].

In this study, 11 generating units are utilized, consisting of eight thermal units, two CHP units and one heat-only unit. Tables 6.1, 6.2, and 6.3 present the data for the thermal, CHP and heat-only generating units, respectively. The load demand and heat demand are presented in Table 6.4. The transmission line loss coefficients for the CHP and thermal-generating units are given in Eqs. (6.11) and (6.12), respectively. The ramp-up and ramp-down rates for the CHP units are given as 70 MW and 50 MW, respectively.

$$B = 10^{-5}x\begin{bmatrix} 3.50 & 1.30 \\ 1.30 & 4.00 \end{bmatrix}\text{per MW} \tag{6.11}$$

$$B = 10^{-5}x\begin{bmatrix} 4.90 & 1.40 & 1.50 & 1.50 & 1.70 & 1.70 & 1.90 & 2.00 \\ 1.40 & 4.50 & 1.60 & 1.60 & 1.50 & 1.50 & 1.80 & 1.80 \\ 1.50 & 1.60 & 3.90 & 1.00 & 1.20 & 1.40 & 1.60 & 1.60 \\ 1.50 & 1.60 & 1.00 & 4.00 & 1.00 & 1.10 & 1.40 & 1.50 \\ 1.70 & 1.50 & 1.20 & 1.00 & 3.60 & 1.30 & 1.40 & 1.50 \\ 1.70 & 1.50 & 1.40 & 1.10 & 1.20 & 3.80 & 1.60 & 1.80 \\ 1.90 & 1.80 & 1.60 & 1.40 & 1.40 & 1.60 & 4.20 & 1.90 \\ 2.00 & 1.80 & 1.60 & 1.50 & 1.50 & 1.80 & 1.90 & 4.40 \end{bmatrix}\text{per MW} \tag{6.12}$$

The optimization model is formulated and solved using the CONPOT solver embedded in AIMMS. The AIMMS code for solving an example of a CHPDED problem is shown below (Table 6.5):

**Table 6.1** Data for thermal-generating units

| Thermal units | $a_j$ | $b_j$ | $c_j$ | $e_j$ | $f_j$ | $\alpha_j$ | $\beta_j$ | $\delta_j$ |
|---|---|---|---|---|---|---|---|---|
| j = 1 | 786.7988 | 38.5397 | 0.1524 | 450 | 0.041 | 103.3908 | 2.4444 | 0.0312 |
| j = 2 | 451.3251 | 46.1591 | 0.1058 | 600 | 0.036 | 103.3908 | 2.4444 | 0.0312 |
| j = 3 | 1049.998 | 40.3965 | 0.028 | 320 | 0.028 | 300.391 | 4.0695 | 0.0509 |
| j = 4 | 1243.531 | 38.3055 | 0.0354 | 260 | 0.052 | 300.391 | 4.0695 | 0.0509 |
| j = 5 | 1356.659 | 38.2704 | 0.0179 | 310 | 0.048 | 320.0006 | 3.8132 | 0.0344 |
| j = 6 | 1450.705 | 36.5104 | 0.0121 | 300 | 0.086 | 330.0056 | 3.9023 | 0.0465 |
| j = 7 | 1455.606 | 39.5804 | 0.109 | 270 | 0.098 | 350.0056 | 3.9524 | 0.0465 |
| j = 8 | 1469.403 | 40.5407 | 0.1295 | 380 | 0.094 | 360.0012 | 3.9864 | 0.047 |

**Table 6.2** Data for CHP-generating units

| CHP units | $a_g$ | $b_g$ | $c_g$ | $d_g$ | $e_g$ | $f_g$ |
|---|---|---|---|---|---|---|
| g = 1 | 2650 | 14.5 | 0.0345 | 4.2 | 0.03 | 0.031 |
| g = 2 | 1250 | 36 | 0.0435 | 0.6 | 0.027 | 0.011 |

**Table 6.3** Data for heat-generating unit

| Heat unit | $a_i$ | $b_i$ | $c_i$ | $\alpha_j$ | $\beta_j$ | $H_{i,\min}^T$ | $H_{i,\max}^T$ |
|---|---|---|---|---|---|---|---|
| i = 1 | 950 | 2.0109 | 0.038 | 0.0008 | 0.001 | 0 | 2695.2 |

**Table 6.4** Data for heat and power demand

| Hour | Load demand | Heat demand |
|---|---|---|
| 1 | 1036 | 390 |
| 2 | 1110 | 400 |
| 3 | 1258 | 410 |
| 4 | 1406 | 420 |
| 5 | 1480 | 440 |
| 6 | 1628 | 450 |
| 7 | 1702 | 450 |
| 8 | 1776 | 455 |
| 9 | 1924 | 460 |
| 10 | 2072 | 460 |
| 11 | 2146 | 470 |
| 12 | 2220 | 480 |
| 13 | 2072 | 470 |
| 14 | 1924 | 460 |
| 15 | 1776 | 450 |

(continued)

**Table 6.4** (continued)

| Hour | Load demand | Heat demand |
|---|---|---|
| 16 | 1554 | 450 |
| 17 | 1480 | 420 |
| 18 | 1628 | 435 |
| 19 | 1776 | 445 |
| 20 | 2072 | 450 |
| 21 | 1924 | 445 |
| 22 | 1628 | 435 |
| 23 | 1332 | 400 |
| 24 | 1184 | 400 |

**Table 6.5** Optimal power flow for CHPDED model

| Hour | $P_1^T$ | $P_2^T$ | $P_3^T$ | $P_4^T$ | $P_5^T$ | $P_6^T$ | $P_7^T$ | $P_8^T$ | $P_1^{CHP}$ | $P_2^{CHP}$ |
|---|---|---|---|---|---|---|---|---|---|---|
| 1 | 150 | 135 | 91.43 | 80.87 | 84.66 | 63.78 | 32.51 | 65.47 | 335.93 | 10.00 |
| 2 | 150 | 135 | 96.56 | 82.30 | 86.03 | 113.78 | 35.90 | 65.79 | 350.23 | 10.00 |
| 3 | 150 | 135 | 104.73 | 84.87 | 88.56 | 163.78 | 85.90 | 66.39 | 377.13 | 21.54 |
| 4 | 150 | 135 | 113.75 | 88.06 | 108.91 | 213.78 | 93.10 | 67.13 | 411.64 | 48.95 |
| 5 | 150 | 135 | 117.85 | 89.60 | 158.91 | 263.78 | 34.88 | 67.49 | 428.09 | 61.23 |
| 6 | 150 | 135 | 123.41 | 91.75 | 208.91 | 313.78 | 39.05 | 67.98 | 450.75 | 79.85 |
| 7 | 150 | 135 | 122.13 | 91.25 | 219.73 | 342.95 | 89.05 | 67.87 | 444.77 | 75.35 |
| 8 | 150 | 135 | 128.87 | 96.08 | 231.91 | 392.95 | 39.82 | 68.47 | 473.13 | 98.87 |
| 9 | 150 | 135 | 127.17 | 146.08 | 281.91 | 415.83 | 89.82 | 68.32 | 463.98 | 92.44 |
| 10 | 150 | 135 | 133.90 | 196.08 | 331.91 | 465.83 | 36.62 | 68.91 | 491.17 | 116.39 |
| 11 | 150 | 135 | 137.60 | 207.10 | 350.67 | 477.60 | 41.38 | 69.23 | 506.26 | 128.67 |
| 12 | 150 | 135 | 140.65 | 208.28 | 351.88 | 478.00 | 91.38 | 69.49 | 518.21 | 138.42 |
| 13 | 150 | 135 | 139.28 | 207.75 | 331.71 | 405.56 | 40.52 | 69.37 | 510.59 | 134.49 |
| 14 | 150 | 135 | 126.59 | 189.55 | 241.71 | 413.85 | 90.52 | 68.27 | 464.45 | 90.39 |
| 15 | 150 | 135 | 118.39 | 99.55 | 198.91 | 463.85 | 93.70 | 67.54 | 428.26 | 62.35 |
| 16 | 150 | 135 | 110.70 | 86.95 | 108.91 | 401.73 | 92.71 | 66.88 | 396.52 | 36.81 |
| 17 | 150 | 135 | 108.54 | 86.18 | 102.59 | 401.47 | 37.91 | 66.70 | 388.70 | 32.27 |
| 18 | 150 | 135 | 121.08 | 92.53 | 152.59 | 341.20 | 87.91 | 67.78 | 440.17 | 72.80 |
| 19 | 150 | 135 | 115.53 | 142.53 | 202.59 | 391.20 | 96.19 | 67.29 | 445.53 | 69.99 |
| 20 | 150 | 135 | 145.53 | 192.53 | 229.34 | 406.38 | 146.19 | 69.88 | 509.40 | 139.99 |
| 21 | 150 | 135 | 129.14 | 192.10 | 222.59 | 404.17 | 95.05 | 68.49 | 472.68 | 100.58 |
| 22 | 150 | 135 | 125.01 | 102.10 | 198.91 | 331.34 | 35.65 | 68.13 | 437.63 | 76.96 |
| 23 | 150 | 135 | 102.98 | 84.29 | 108.91 | 256.30 | 33.42 | 66.26 | 390.38 | 26.96 |
| 24 | 150 | 135 | 98.57 | 82.89 | 86.61 | 183.57 | 33.04 | 65.93 | 356.28 | 10.00 |

```
Model Main_Economic_Dispatch_ten_bus_ded {
  Set conventional {
    Index: i, k1;
  }
  Set heat {
    Index: h;
  }
  Set chp {
    Index: p, k2;
  }
  Set Time {
    Index: t;
  }
  Parameter pmin {
    IndexDomain: i;
  }
  Parameter pmax {
    IndexDomain: i;
  }
  Parameter a {
    IndexDomain: i;
  }
  Parameter b {
    IndexDomain: i;
  }
  Parameter c {
    IndexDomain: i;
  }
  Parameter e {
    IndexDomain: i;
  }
  Parameter f {
    IndexDomain: i;
  }
  Parameter ma {
    IndexDomain: i;
  }
  Parameter mb {
    IndexDomain: i;
  }
  Parameter mc {
    IndexDomain: i;
  }
  Parameter mem {
    IndexDomain: i;
  }
  Parameter mf {
    IndexDomain: i;
  }
  Parameter w {
    Definition: 0.5;
  }
  Parameter aa {
    IndexDomain: p;
  }
  Parameter bb {
    IndexDomain: p;
  }
  Parameter cc {
    IndexDomain: p;
  }
```

```
Parameter dd {
   IndexDomain: p;
}
Parameter ee {
   IndexDomain: p;
}
Parameter ff {
   IndexDomain: p;
}
Parameter maa {
   IndexDomain: p;
}
Parameter mbb {
   IndexDomain: p;
}
Parameter aaa {
   IndexDomain: h;
}
Parameter ccc {
   IndexDomain: h;
}
Parameter bbb {
   IndexDomain: h;
}
Parameter maaa {
   Definition: 0.0008;
}
Parameter mbbb {
   Definition: 0.0010;
}
Parameter dr {
   IndexDomain: i;
}
Parameter drch {
   IndexDomain: p;
}
Parameter urch {
   IndexDomain: p;
}
Parameter ur {
   IndexDomain: i;
}
Parameter d {
   IndexDomain: t;
}
Parameter he {
   IndexDomain: t;
}
Parameter bma1 {
   IndexDomain: (i,k1);
}
Parameter bma2 {
   IndexDomain: (p,k2);
}
Parameter loss1 {
   IndexDomain: t;
   Definition: sum[i,sum[k1, pc(i,t)*bma1(i,k1)*pc(k1,t)]];
}
Parameter loss2 {
   IndexDomain: t;
   Definition: sum[p,sum[k2, pch(p,t)*bma2(p,k2)*pch(k2,t)]];
}
```

```
Variable pc {
   IndexDomain: (i,t);
   Range: nonnegative;
}
Variable pch {
   IndexDomain: (p,t);
   Range: nonnegative;
}
Variable hch {
   IndexDomain: (p,t);
   Range: nonnegative;
}
Variable hh {
   IndexDomain: (h,t);
   Range: (0, 2695.2);
}
Variable Total {
   Range: free;
   Definition: {
      (sum[(i,t), a(i)+(b(i)*pc(i,t))+(c(i)*pc(i,t)*pc(i,t))+(e(i)*sin(f(i)*(pmin(i)-pc(i,t))))]+
      sum[(p,t),
aa(p)+(bb(p)*pch(p,t))+(cc(p)*pch(p,t)*pch(p,t))+(dd(p)*hch(p,t))+(ee(p)*hch(p,t)*hch(p,t))+(ff(p)*pch(p,t)*hch(p,
t))]+
      sum[(h,t), aaa(h)+(bbb(h)*hh(h,t))+(ccc(h)*hh(h,t)*hh(h,t))])
   }
   Comment: {
      "(sum[(i,t), a(i)+(b(i)*pc(i,t))+(c(i)*pc(i,t)*pc(i,t))+(e(i)*sin(f(i)*(pmin(i)-pc(i,t))))]+
      sum[(p,t),
aa(p)+(bb(p)*pch(p,t))+(cc(p)*pch(p,t)*pch(p,t))+(dd(p)*hch(p,t))+(ee(p)*hch(p,t)*hch(p,t))+(ff(p)*pch(p,t)*hch(p,
t))]+
      sum[(h,t), aaa(h)+(bbb(h)*hh(h,t))+(ccc(h)*hh(h,t)*hh(h,t))])"
   }
}
Parameter pchmin {
   IndexDomain: p;
}
Parameter hchmin {
   IndexDomain: p;
}
Parameter hchmax {
   IndexDomain: p;
}
Parameter pchmax {
   IndexDomain: p;
}
Parameter hhmin {
   IndexDomain: h;
   Definition: 0;
}
Parameter hhmax {
   IndexDomain: h;
   Definition: 2695.2;
}
Constraint sup {
   IndexDomain: t;
   Definition: sum[i, pc(i,t)]+sum[p, pch(p,t)]=d(t)+loss1(t)+loss2(t);
   Comment: "sum[i, pc(i,t)]+sum[p, pch(p,t)]=d(t)+loss1(t)+loss2(t)";
}
Constraint supp {
   IndexDomain: t;
   Definition: sum[h, hh(h,t)]+sum[p, hch(p,t)]=he(t);
}
```

```
Constraint sup1 {
   IndexDomain: (i,t);
   Definition: pc(i,t)<=pmax(i);
}
Constraint sup2 {
   IndexDomain: (i,t);
   Definition: pc(i,t)>=pmin(i);
}
Constraint sup3 {
   IndexDomain: (i,t) | 1<= ord(t) <=23;
   Definition: pc(i,t)-pc(i,t+1)<=dr(i);
}
Constraint sup4 {
   IndexDomain: (i,t) | 1<= ord(t) <=23;
   Definition: pc(i,t+1)-pc(i,t)<=ur(i);
}
Constraint sup5 {
   IndexDomain: (p,t);
   Definition: pchmin(p)<=pch(p,t);
   Comment: "pchmin(p)<=pch(p,t)<=pchmax(p)";
}
Constraint sup6 {
   IndexDomain: (p,t);
   Definition: 1=1;
   Comment: "pch(p,t)<=pchmax(p)";
}
Constraint sup7 {
   IndexDomain: (h,t);
   Definition: 1=1;
}
Constraint sup8 {
   IndexDomain: (p,t)| 1<= ord(t) <=23;
   Definition: -drch(p)<=pch(p,t+1)-pch(p,t);
   Comment: "-drch(p)<=pch(p,t+1)-pch(p,t)<=urch(p)";
}
Constraint sup9 {
   IndexDomain: (p,t)| 1<= ord(t) <=23;
   Definition: pch(p,t+1)-pch(p,t)<=urch(p);
}
MathematicalProgram LeastEconDis {
   Objective: Total;
   Direction: minimize;
   Constraints: AllConstraints;
   Variables: AllVariables;
   Type: Automatic;
}
Procedure MainInitialization;
Procedure MainExecution {
   Body: {
      solve LeastEconDis;
   }
}
```

Tables 6.5 and 6.6 present the optimal power and heat flow for the operational schedule of the 11 generating units. The simulated results show variation in the hourly optimal power dispatch of the thermal, CHP and heat-only generators.

**Table 6.6** Optimal heat flow and losses for CHPDED model

| Hour | $H_1^{\mathrm{CHP}}$ | $H_2^{\mathrm{CHP}}$ | $H_1^H$ | $\mathrm{Loss}_k$ |
|---|---|---|---|---|
| 1 | 143.01 | 95 | 151.99 | 13.65 |
| 2 | 148.85 | 95 | 156.15 | 15.60 |
| 3 | 153.72 | 95 | 161.28 | 19.89 |
| 4 | 157.25 | 95 | 167.75 | 24.33 |
| 5 | 167.90 | 95 | 177.10 | 26.82 |
| 6 | 172.17 | 95 | 182.83 | 32.48 |
| 7 | 172.55 | 95 | 182.45 | 36.11 |
| 8 | 173.49 | 95 | 186.51 | 39.10 |
| 9 | 176.95 | 95 | 188.05 | 46.54 |
| 10 | 174.93 | 95 | 190.07 | 53.80 |
| 11 | 179.74 | 95 | 195.26 | 57.51 |
| 12 | 184.76 | 95 | 200.24 | 61.30 |
| 13 | 179.24 | 95 | 195.76 | 52.27 |
| 14 | 177.13 | 95 | 187.87 | 46.32 |
| 15 | 173.65 | 95 | 181.35 | 41.55 |
| 16 | 175.82 | 95 | 179.18 | 32.22 |
| 17 | 158.66 | 95 | 166.34 | 29.36 |
| 18 | 164.00 | 95 | 176.00 | 33.06 |
| 19 | 170.08 | 95 | 179.92 | 39.87 |
| 20 | 167.08 | 95 | 187.92 | 52.24 |
| 21 | 167.50 | 95 | 182.50 | 45.81 |
| 22 | 163.65 | 95 | 176.35 | 32.73 |
| 23 | 147.42 | 95 | 157.58 | 22.50 |
| 24 | 148.85 | 95 | 156.15 | 17.90 |

## 6.3 Combined Heat and Power Pure Dynamic Emissions Dispatch

Recently, there has been greater environmental awareness of emissions such as $CO_2$, $SO_2$ and $NO_x$ from electricity-generating units. The preferred solution approach to solve the problem of minimizing emission is emission dispatching [4]. The most effective methods of integrating emission into optimal dispatching is considering the emission as either an objective function or a constraint.

In this study, the emission from the CHP-generating units is regarded as the main objective function and this is termed CHP pure DED (CHPPDED).

The total emissions (for thermal, CHP and heat units) are given by:

$$\mathrm{Min}[E(P,H)] = \mathrm{Min}\left[\sum_{k=1}^{k}\sum_{j=1}^{J} E_j^T\left(P_{j,k}^T\right) + \sum_{k=1}^{K}\sum_{g=1}^{G} E_g^{\mathrm{CHP}}\left(P_{g,k}^{\mathrm{CHP}}\right) + \sum_{k=1}^{K}\sum_{i=1}^{I} E_i^H\left(H_{i,k}^H\right)\right] \tag{6.13}$$

subject to the following constraints:

$$\sum_{j=1}^{J} P_{j,k}^{T} + \sum_{g=1}^{G} P_{g,k}^{\mathrm{CHP}} = D_k + \mathrm{Loss}_k \tag{6.14}$$

$$\sum_{g=1}^{G} H_{g,k}^{\mathrm{CHP}} + \sum_{i=1}^{I} H_{i,k}^{H} = \mathrm{HD}_k + \mathrm{Loss}_k \tag{6.15}$$

$$P_{j,\min}^{T} \le P_{j,k}^{T} \le P_{j,\max}^{T} \tag{6.16}$$

$$P_{g,\min}^{\mathrm{CHP}}\left(H_{g,k}^{\mathrm{CHP}}\right) \le P_{g,k}^{\mathrm{CHP}} \le P_{g,\max}^{\mathrm{CHP}}\left(H_{g,k}^{\mathrm{CHP}}\right) \tag{6.17}$$

$$H_{g,\min}^{\mathrm{CHP}}\left(P_{g,k}^{\mathrm{CHP}}\right) \le H_{g,k}^{\mathrm{CHP}} \le H_{g,\max}^{\mathrm{CHP}}\left(P_{g,k}^{\mathrm{CHP}}\right) \tag{6.18}$$

$$H_{g,\min}^{H} \le H_{g,k}^{H} \le H_{g,\max}^{H} \tag{6.19}$$

$$-\mathrm{DR}_j^T \le P_{j,k+1}^{T} - P_{j,k}^{T} \le \mathrm{UR}_j^T \tag{6.20}$$

$$-\mathrm{DR}_g^{\mathrm{CHP}} \le P_{g,t+1}^{\mathrm{CHP}} - P_{g,t}^{\mathrm{CHP}} \le \mathrm{UR}_g^{\mathrm{CHP}} \tag{6.21}$$

$$\mathrm{Loss}_k = \sum_{j=1}^{J}\sum_{z=1}^{Z} P_{j,k} B_{j,z} P_{z,k}. \tag{6.22}$$

The desired variable result to be obtained by the optimization model is $P_{j,k}^{T}$, $P_{g,k}^{\mathrm{CHP}}$, $H_{g,k}^{\mathrm{CHP}}$, $H_{i,k}^{H}$ and $\mathrm{Loss}_k$.

Equation (6.14) is a power balance constraint, which compels the total output power from both thermal and CHP-generating units to satisfy the load demand and line losses. Constraint (6.15) is a heat balance constraint and states that the heat output from both heat and CHP units must meet the heat demand. Constraint (6.16) gives the generation limits for thermal-generating units and states that output power from the thermal units must not exceed the allowed limits. The next constraint (6.17) limits power produced by CHP units within the allowable limits. Constraint (6.18) limits the heat produced by CHP units within the allowable limits. Constraint (6.19) ensures that heat produced by heat-only units is within the allowable limits. Constraints (6.20) and (6.21) are the generator ramp rate limits for the thermal- and CHP-generating units. Equation (6.22) is the transmission line losses constraint and the line losses are determined by the B-coefficient method [4].

The data for thermal, CHP and heat-only generating units are presented in Tables 6.1, 6.2, and 6.3. Similarly, all necessary data are presented in Sect. (6.2). The optimization model is formulated and solved using the CONPOT solver embedded in AIMMS. The AIMMS code for solving an example of the CHPPDED problem is shown below:

```
Model Main_Economic_Dispatch_ten_bus_ded {
  Set conventional {
    Index: i, k1;
  }
  Set heat {
    Index: h;
  }
  Set chp {
    Index: p, k2;
  }
  Set Time {
    Index: t;
  }
  Parameter pmin {
    IndexDomain: i;
  }
  Parameter pmax {
    IndexDomain: i;
  }
  Parameter a {
    IndexDomain: i;
  }
  Parameter b {
    IndexDomain: i;
  }
  Parameter c {
    IndexDomain: i;
  }
  Parameter e {
    IndexDomain: i;
  }
  Parameter f {
    IndexDomain: i;
  }
  Parameter ma {
    IndexDomain: i;
  }
  Parameter mb {
    IndexDomain: i;
  }
  Parameter mc {
    IndexDomain: i;
  }
  Parameter mem {
    IndexDomain: i;
  }
  Parameter mf {
    IndexDomain: i;
  }
  Parameter w {
    Definition: 0.5;
  }
  Parameter aa {
    IndexDomain: p;
  }
  Parameter bb {
    IndexDomain: p;
  }
  Parameter cc {
    IndexDomain: p;
  }
  Parameter dd {
    IndexDomain: p;
  }
  Parameter ee {
    IndexDomain: p;
  }
  Parameter ff {
```

```
}
Parameter mbb {
   IndexDomain: p;
}
Parameter aaa {
   IndexDomain: h;
}
Parameter ccc {
   IndexDomain: h;
}
Parameter bbb {
   IndexDomain: h;
}
Parameter maaa {
   Definition: 0.0008;
}
Parameter mbbb {
   Definition: 0.0010;
}
Parameter dr {
   IndexDomain: i;
}
Parameter drch {
   IndexDomain: p;
}
Parameter urch {
   IndexDomain: p;
}
Parameter ur {
   IndexDomain: i;
}
Parameter d {
   IndexDomain: t;
}
Parameter he {
   IndexDomain: t;
}
Parameter bma1 {
   IndexDomain: (i,k1);
}
Parameter bma2 {
   IndexDomain: (p,k2);
}
Parameter loss1 {
   IndexDomain: t;
   Definition: sum[i,sum[k1, pc(i,t)*bma1(i,k1)*pc(k1,t)]];
}
Parameter loss2 {
   IndexDomain: t;
   Definition: sum[p,sum[k2, pch(p,t)*bma2(p,k2)*pch(k2,t)]];
}
Variable pc {
   IndexDomain: (i,t);
   Range: nonnegative;
}
Variable pch {
   IndexDomain: (p,t);
   Range: nonnegative;
}
Variable hch {
   IndexDomain: (p,t);
   Range: nonnegative;
}
Variable hh {
   IndexDomain: (h,t);
   Range: (0, 2695.2);
}
Variable Total {
   Range: free;
```

```
    Definition: {
      (sum[(i,t), ma(i)+(mb(i)*pc(i,t))+(mc(i)*pc(i,t)*pc(i,t))+(mem(i)*exp(mf(i)*(pc(i,t))))]+
      sum[(p,t), (maa(p)+(mbb(p)))*pch(p,t)]+
      sum[(h,t), ((maaa+mbbb)*hh(h,t))])
    }
    Comment: {
      "(sum[(i,t), a(i)+(b(i)*pc(i,t))+(c(i)*pc(i,t)*pc(i,t))+(e(i)*sin(f(i)*(pmin(i)-pc(i,t))))]+
      sum[(p,t),
aa(p)+(bb(p)*pch(p,t))+(cc(p)*pch(p,t)*pch(p,t))+(dd(p)*hch(p,t))+(ee(p)*hch(p,t)*hch(p,t))+(ff(p)*pch(p,t)*hch(p,t))]+
      sum[(h,t), aaa(h)+(bbb(h)*hh(h,t))+(ccc(h)*hh(h,t)*hh(h,t))])"
    }
  }
  Parameter pchmin {
    IndexDomain: p;
  }
  Parameter hchmin {
    IndexDomain: p;
  }
  Parameter hchmax {
    IndexDomain: p;
  }
  Parameter pchmax {
    IndexDomain: p;
  }
  Parameter hhmin {
    IndexDomain: h;
    Definition: 0;
  }
  Parameter hhmax {
    IndexDomain: h;
    Definition: 2695.2;
  }
  Constraint sup {
    IndexDomain: t;
    Definition: sum[i, pc(i,t)]+sum[p, pch(p,t)]=d(t)+loss1(t)+loss2(t);
    Comment: "sum[i, pc(i,t)]+sum[p, pch(p,t)]=d(t)+loss1(t)+loss2(t)";
  }
  Constraint supp {
    IndexDomain: t;
    Definition: sum[h, hh(h,t)]+sum[p, hch(p,t)]=he(t);
  }
  Constraint sup1 {
    IndexDomain: (i,t);
    Definition: pc(i,t)<=pmax(i);
  }
  Constraint sup2 {
    IndexDomain: (i,t);
    Definition: pc(i,t)>=pmin(i);
  }
  Constraint sup3 {
    IndexDomain: (i,t) | 1<= ord(t) <=23;
    Definition: pc(i,t)-pc(i,t+1)<=dr(i);
  }
  Constraint sup4 {
    IndexDomain: (i,t) | 1<= ord(t) <=23;
    Definition: pc(i,t+1)-pc(i,t)<=ur(i);
  }
  Constraint sup5 {
    IndexDomain: (p,t);
    Definition: pchmin(p)<=pch(p,t);
    Comment: "pchmin(p)<=pch(p,t)<=pchmax(p)";
  }
  Constraint sup6 {
    IndexDomain: (p,t);
    Definition: 1=1;
    Comment: "pch(p,t)<=pchmax(p)";
  }
  Constraint sup7 {
```

```
    IndexDomain: (h,t);
    Definition: 1=1;
}
Constraint sup8 {
    IndexDomain: (p,t)| 1<= ord(t) <=23;
    Definition: -drch(p)<=pch(p,t+1)-pch(p,t);
    Comment: "-drch(p)<=pch(p,t+1)-pch(p,t)<=urch(p)";
}
Constraint sup9 {
    IndexDomain: (p,t)| 1<= ord(t) <=23;
    Definition: pch(p,t+1)-pch(p,t)<=urch(p);
}
MathematicalProgram LeastEconDis {
    Objective: Total;
    Direction: minimize;
    Constraints: AllConstraints;
    Variables: AllVariables;
    Type: Automatic;
}
Procedure MainInitialization;
Procedure MainExecution {
    Body: {
        solve LeastEconDis;
    }
}
```

Tables 6.7 and 6.8 depict the optimal power and heat flow for the operational schedule of the generating units in a CHPPDED model. The simulated results show variation in the hourly optimal power dispatch of the thermal, CHP and heat-only generators.

**Table 6.7** Optimal power flow for CHPPDED model

| Hour | $P_1^T$ | $P_2^T$ | $P_3^T$ | $P_4^T$ | $P_5^T$ | $P_6^T$ | $P_7^T$ | $P_8^T$ | $P_1^{CHP}$ | $P_2^{CHP}$ |
|---|---|---|---|---|---|---|---|---|---|---|
| 1 | 150 | 135 | 123 | 90 | 107 | 55 | 55 | 55 | 270.56 | 10.01 |
| 2 | 150 | 135 | 123 | 90 | 107 | 55 | 105 | 105 | 235.14 | 21.90 |
| 3 | 150 | 135 | 123 | 90 | 127 | 106.52 | 95.98 | 55 | 305.14 | 91.90 |
| 4 | 150 | 135 | 123 | 70 | 97 | 133.28 | 132.74 | 55 | 375.14 | 161.90 |
| 5 | 150 | 135 | 123 | 100 | 127 | 90 | 90 | 55 | 445.14 | 196.33 |
| 6 | 150 | 135 | 123 | 100 | 107 | 127.90 | 137.36 | 105 | 415.14 | 266.33 |
| 7 | 150 | 135 | 123 | 90 | 127 | 120 | 120 | 105 | 485.14 | 291.06 |
| 8 | 150 | 135 | 123 | 90 | 97 | 100 | 120 | 255 | 444.05 | 311.35 |
| 9 | 150 | 135 | 123 | 70 | 97 | 129.02 | 128.48 | 255 | 514.05 | 381.35 |
| 10 | 150 | 135 | 123 | 70 | 97 | 138.28 | 137.74 | 255 | 584.05 | 451.35 |
| 11 | 150 | 135 | 123 | 100 | 137.91 | 214.07 | 153.54 | 192.50 | 518.25 | 403.22 |
| 12 | 157.67 | 157.67 | 120.68 | 100.68 | 153.10 | 200.11 | 189.57 | 288.26 | 568.25 | 353.22 |
| 13 | 150.00 | 148.15 | 74.85 | 174.85 | 114.48 | 183.73 | 183.19 | 281.94 | 518.25 | 303.22 |
| 14 | 150 | 136.22 | 123 | 97.53 | 123.65 | 175.72 | 175.18 | 274.02 | 468.25 | 253.22 |
| 15 | 150 | 135 | 103 | 90.00 | 129.19 | 165.03 | 164.49 | 263.44 | 418.25 | 203.22 |
| 16 | 150 | 135 | 103 | 90 | 97 | 120 | 120 | 155 | 468.25 | 153.22 |

(continued)

**Table 6.7** (continued)

| Hour | $P_1^T$ | $P_2^T$ | $P_3^T$ | $P_4^T$ | $P_5^T$ | $P_6^T$ | $P_7^T$ | $P_8^T$ | $P_1^{CHP}$ | $P_2^{CHP}$ |
|---|---|---|---|---|---|---|---|---|---|---|
| 17 | 150 | 135 | 93 | 60 | 107 | 150 | 120 | 155 | 418.88 | 124.56 |
| 18 | 150 | 135 | 83 | 110 | 107 | 127.88 | 117.34 | 155 | 488.88 | 194.56 |
| 19 | 150 | 135 | 123 | 110 | 150.96 | 129.83 | 129.29 | 158.03 | 470.87 | 263.54 |
| 20 | 212.82 | 205.00 | 134.49 | 150.00 | 180.96 | 159.83 | 219.29 | 228.03 | 420.87 | 213.54 |
| 21 | 184.96 | 184.96 | 197.41 | 197.41 | 147.86 | 158.43 | 157.89 | 206.38 | 370.87 | 163.54 |
| 22 | 151.13 | 151.13 | 126.67 | 126.67 | 117.17 | 185.73 | 185.19 | 183.92 | 320.87 | 113.54 |
| 23 | 150 | 135 | 73 | 100 | 61.24 | 104.34 | 143.81 | 155 | 370.87 | 63.54 |
| 24 | 150 | 135 | 123 | 100 | 97 | 120 | 90 | 155 | 220.87 | 13.54 |

**Table 6.8** Optimal heat flow for CHPPDED model

| Hour | $H_1^{CHP}$ | $H_2^{CHP}$ | $H_1^H$ |
|---|---|---|---|
| 1 | 42.63 | 347.37 | 14.57 |
| 2 | 16.78 | 383.22 | 17.04 |
| 3 | 1.56 | 408.44 | 21.53 |
| 4 | 1.60 | 418.40 | 27.05 |
| 5 | 1.67 | 438.33 | 31.47 |
| 6 | 1.71 | 448.29 | 38.73 |
| 7 | 1.71 | 448.29 | 44.20 |
| 8 | 1.73 | 453.27 | 49.40 |
| 9 | 1.75 | 458.25 | 58.90 |
| 10 | 1.75 | 458.25 | 69.42 |
| 11 | 1.78 | 468.22 | 71.49 |
| 12 | 1.82 | 478.18 | 69.22 |
| 13 | 1.78 | 468.22 | 60.67 |
| 14 | 1.75 | 458.25 | 52.80 |
| 15 | 1.71 | 448.29 | 45.61 |
| 16 | 1.71 | 448.29 | 37.47 |
| 17 | 1.60 | 418.40 | 33.43 |
| 18 | 1.65 | 433.35 | 40.65 |
| 19 | 1.69 | 443.31 | 44.52 |
| 20 | 1.71 | 448.29 | 52.82 |
| 21 | 1.69 | 443.31 | 45.70 |
| 22 | 1.65 | 433.35 | 34.01 |
| 23 | 1.53 | 398.47 | 24.79 |
| 24 | 1.53 | 398.47 | 20.41 |

## 6.4 Combined Heat and Power Dynamic Economic Emissions Dispatch

The DEED is a multi-objective optimization problem encompassing two different objectives [5]. These objectives are to minimize fuel cost and emissions from the generating units simultaneously. The main aim of CHPDEED is to determine the optimal power and heat output of the thermal, CHP and heat-only generators under technical and economic constraints.

The individual fuel cost and emissions objective functions of all three types of generating units (thermal, CHP and heat) are detailed below.

$$C(P,H) = \sum_{k=1}^{k}\sum_{j=1}^{J} C_j^T\left(P_{j,k}^T\right) + \sum_{k=1}^{K}\sum_{g=1}^{G} C_g^{\mathrm{CHP}}\left(P_{g,k}^{\mathrm{CHP}}, H_{g,k}^{\mathrm{CHP}}\right) + \sum_{k=1}^{K}\sum_{i=1}^{I} C_i^H\left(H_{i,k}^H\right) \tag{6.23}$$

$$E(P,H) = \sum_{k=1}^{k}\sum_{j=1}^{J} E_j^T\left(P_{j,k}^T\right) + \sum_{k=1}^{K}\sum_{g=1}^{G} E_g^{\mathrm{CHP}}\left(P_{g,k}^{\mathrm{CHP}}\right) + \sum_{k=1}^{K}\sum_{i=1}^{I} E_i^H\left(H_{i,k}^H\right) \tag{6.24}$$

$$\mathrm{Min}[u_1(C(P,H)) + u_2(E(P,H))] \tag{6.25}$$

subject to the following constraints:

$$\sum_{j=1}^{J} P_{j,k}^T + \sum_{g=1}^{G} P_{g,k}^{\mathrm{CHP}} = D_k + \mathrm{Loss}_k \tag{6.26}$$

$$\sum_{g=1}^{G} H_{g,k}^{\mathrm{CHP}} + \sum_{i=1}^{I} H_{i,k}^H = \mathrm{HD}_k + \mathrm{Loss}_k \tag{6.27}$$

$$P_{j,\min}^T \le P_{j,k}^T \le P_{j,\max}^T \tag{6.28}$$

$$P_{g,\min}^{\mathrm{CHP}}\left(H_{g,k}^{\mathrm{CHP}}\right) \le P_{g,k}^{\mathrm{CHP}} \le P_{g,\max}^{\mathrm{CHP}}\left(H_{g,k}^{\mathrm{CHP}}\right) \tag{6.29}$$

$$H_{g,\min}^{\mathrm{CHP}}\left(P_{g,k}^{\mathrm{CHP}}\right) \le H_{g,k}^{\mathrm{CHP}} \le H_{g,\max}^{\mathrm{CHP}}\left(P_{g,k}^{\mathrm{CHP}}\right) \tag{6.30}$$

$$H^{H}_{g,\min} \leq H^{H}_{g,k} \leq H^{H}_{g,\max} \tag{6.31}$$

$$-DR^{T}_{j} \leq P^{T}_{j,k+1} - P^{T}_{j,k} \leq UR^{T}_{j} \tag{6.32}$$

$$-DR^{\mathrm{CHP}}_{g} \leq P^{\mathrm{CHP}}_{g,t+1} - P^{\mathrm{CHP}}_{g,t} \leq UR^{\mathrm{CHP}}_{g} \tag{6.33}$$

$$\mathrm{Loss}_{k} = \sum_{j=1}^{J}\sum_{z=1}^{Z} P_{j,k}B_{j,z}P_{z,k}.s \tag{6.34}$$

The desired variable result to be obtained by the optimization model is $P^{T}_{j,k}$, $P^{\mathrm{CHP}}_{g,k}$, $H^{\mathrm{CHP}}_{g,k}$, $H^{H}_{i,k}$ and $\mathrm{Loss}_{k}$.

Equation (6.23) is the economic objective function that depicts the total fuel cost of the thermal, CHP and heat-only generators. Equation (6.24) is the total emissions from the generating units as an objective function. To combine the two conflicting objectives, a weighting factor is used, as shown in Eq. (6.25) [6].

Equation (6.26) is a power balance constraint, which compels the total output power from both thermal- and CHP-generating units to satisfy the load demand and line losses. Constraint (6.27) is the heat balance constraint, which states that the heat output from both heat and CHP units must meet the heat demand. Constraint (6.28) gives the generation limits for thermal-generating units and states that output power from the thermal units must not exceed the allowed limits. The next constraint (6.29) limits power produced by CHP units within the allowable limits. Constraint (6.30) limits the heat produced by CHP units within the allowable limits. Constraint (6.31) ensures that heat produced by heat-only units is within the allowable limits. Constraints (6.32) and (6.33) are the generator ramp rate limits for the thermal- and CHP-generating units. Equation (6.34) is the transmission line losses constraint and the line losses are determined by the *B*-coefficient method [4].

The data for thermal, CHP and heat-only generating units are presented in Tables 6.1, 6.2 and 6.3. Similarly, all necessary data are presented in Sect. (6.2). The optimization model is formulated and solved using the CONPOT solver embedded in AIMMS. The AIMMS code for solving an example of the CHPDEED problem is shown below:

```
Model Main_Economic_Dispatch_ten_bus_ded {
  Set conventional {
    Index: i, k1;
  }
  Set heat {
    Index: h;
  }
  Set chp {
    Index: p, k2;
  }
  Set Time {
    Index: t;
  }
  Parameter pmin {
    IndexDomain: i;
  }
  Parameter pmax {
    IndexDomain: i;
  }
  Parameter a {
    IndexDomain: i;
  }
  Parameter b {
    IndexDomain: i;
  }
  Parameter c {
    IndexDomain: i;
  }
  Parameter e {
    IndexDomain: i;
  }
  Parameter f {
    IndexDomain: i;
  }
  Parameter ma {
    IndexDomain: i;
  }
  Parameter mb {
    IndexDomain: i;
  }
  Parameter mc {
    IndexDomain: i;
  }
  Parameter mem {
    IndexDomain: i;
  }
  Parameter mf {
    IndexDomain: i;
  }
  Parameter w {
    Definition: 0.5;
  }
  Parameter aa {
    IndexDomain: p;
  }
  Parameter bb {
    IndexDomain: p;
  }
  Parameter cc {
    IndexDomain: p;
  }
  Parameter dd {
    IndexDomain: p;
  }
  Parameter ee {
    IndexDomain: p;
  }
  Parameter ff {
    IndexDomain: p;
```

```
}
Parameter maa {
  IndexDomain: p;
}
Parameter mbb {
  IndexDomain: p;
}
Parameter aaa {
  IndexDomain: h;
}
Parameter ccc {
  IndexDomain: h;
}
Parameter bbb {
  IndexDomain: h;
}
Parameter maaa {
  Definition: 0.0008;
}
Parameter mbbb {
  Definition: 0.0010;
}
Parameter dr {
  IndexDomain: i;
}
Parameter drch {
  IndexDomain: p;
}
Parameter urch {
  IndexDomain: p;
}
Parameter ur {
  IndexDomain: i;
}
Parameter d {
  IndexDomain: t;
}
Parameter he {
  IndexDomain: t;
}
Parameter bma1 {
  IndexDomain: (i,k1);
}
Parameter bma2 {
  IndexDomain: (p,k2);
}
Parameter loss1 {
  IndexDomain: t;
  Definition: sum[i,sum[k1, pc(i,t)*bma1(i,k1)*pc(k1,t)]];
}
Parameter loss2 {
  IndexDomain: t;
  Definition: sum[p,sum[k2, pch(p,t)*bma2(p,k2)*pch(k2,t)]];
}
Variable pc {
  IndexDomain: (i,t);
  Range: nonnegative;
}
Variable pch {
  IndexDomain: (p,t);
  Range: nonnegative;
}
Variable hch {
  IndexDomain: (p,t);
  Range: nonnegative;
}
Variable hh {
  IndexDomain: (h,t);
```

```
    Range: (0, 2695.2);
  }
  Variable Total {
    Range: free;
    Definition: {
      (w*(sum[(i,t), a(i)+(b(i)*pc(i,t))+(c(i)*pc(i,t)*pc(i,t))+(e(i)*sin(f(i)*(pmin(i)-pc(i,t))))]+
      sum[(p,t),
aa(p)+(bb(p)*pch(p,t))+(cc(p)*pch(p,t)*pch(p,t))+(dd(p)*hch(p,t))+(ee(p)*hch(p,t)*hch(p,t))+(ff(p)*pch(p,t)*hch(p,t))]+
      sum[(h,t), aaa(h)+(bbb(h)*hh(h,t))+(ccc(h)*hh(h,t)*hh(h,t))]))
      +
      ((1-w)*(sum[(i,t), ma(i)+(mb(i)*pc(i,t))+(mc(i)*pc(i,t)*pc(i,t))+(mem(i)*exp(mf(i)*(pc(i,t))))]+
      sum[(p,t), (maa(p)+(mbb(p)))*pch(p,t)]+
      sum[(h,t), ((maaa+mbbb)*hh(h,t))]))
    }
    Comment: {
      "(sum[(i,t), a(i)+(b(i)*pc(i,t))+(c(i)*pc(i,t)*pc(i,t))+(e(i)*sin(f(i)*(pmin(i)-pc(i,t))))]+
      sum[(p,t),
aa(p)+(bb(p)*pch(p,t))+(cc(p)*pch(p,t)*pch(p,t))+(dd(p)*hch(p,t))+(ee(p)*hch(p,t)*hch(p,t))+(ff(p)*pch(p,t)*hch(p,t))]+
      sum[(h,t), aaa(h)+(bbb(h)*hh(h,t))+(ccc(h)*hh(h,t)*hh(h,t))])"
    }
  }
  Parameter pchmin {
    IndexDomain: p;
  }
  Parameter hchmin {
    IndexDomain: p;
  }
  Parameter hchmax {
    IndexDomain: p;
  }
  Parameter pchmax {
    IndexDomain: p;
  }
  Parameter hhmin {
    IndexDomain: h;
    Definition: 0;
  }
  Parameter hhmax {
    IndexDomain: h;
    Definition: 2695.2;
  }
  Constraint sup {
    IndexDomain: t;
    Definition: sum[i, pc(i,t)]+sum[p, pch(p,t)]=d(t)+loss1(t)+loss2(t);
    Comment: "sum[i, pc(i,t)]+sum[p, pch(p,t)]=d(t)+loss1(t)+loss2(t)";
  }
  Constraint supp {
    IndexDomain: t;
    Definition: sum[h, hh(h,t)]+sum[p, hch(p,t)]=he(t);
  }
  Constraint sup1 {
    IndexDomain: (i,t);
    Definition: pc(i,t)<=pmax(i);
  }
  Constraint sup2 {
    IndexDomain: (i,t);
    Definition: pc(i,t)>=pmin(i);
  }
  Constraint sup3 {
    IndexDomain: (i,t) | 1<= ord(t) <=23;
    Definition: pc(i,t)-pc(i,t+1)<=dr(i);
  }
  Constraint sup4 {
    IndexDomain: (i,t) | 1<= ord(t) <=23;
    Definition: pc(i,t+1)-pc(i,t)<=ur(i);
  }
  Constraint sup5 {
    IndexDomain: (p,t);
    Definition: pchmin(p)<=pch(p,t);
```

```
    Comment: "pchmin(p)<=pch(p,t)<=pchmax(p)";
}
Constraint sup6 {
    IndexDomain: (p,t);
    Definition: 1=1;
    Comment: "pch(p,t)<=pchmax(p)";
}
Constraint sup7 {
    IndexDomain: (h,t);
    Definition: 1=1;
}
Constraint sup8 {
    IndexDomain: (p,t)| 1<= ord(t) <=23;
    Definition: -drch(p)<=pch(p,t+1)-pch(p,t);
    Comment: "-drch(p)<=pch(p,t+1)-pch(p,t)<=urch(p)";
}
Constraint sup9 {
    IndexDomain: (p,t)| 1<= ord(t) <=23;
    Definition: pch(p,t+1)-pch(p,t)<=urch(p);
}
MathematicalProgram LeastEconDis {
    Objective: Total;
    Direction: minimize;
    Constraints: AllConstraints;
    Variables: AllVariables;
    Type: Automatic;
}
Procedure MainInitialization;
Procedure MainExecution {
    Body: {
        solve LeastEconDis;
    }
```

Tables 6.9 and 6.10 give the numerical simulation results obtained from the optimization problem. The results show variation in the hourly power and heat flow from the thermal, CHP and heat-only generators.

**Table 6.9** Optimal power flow for CHPDEED model

| Hour | $P_1^T$ | $P_2^T$ | $P_3^T$ | $P_4^T$ | $P_5^T$ | $P_6^T$ | $P_7^T$ | $P_8^T$ | $P_1^{CHP}$ | $P_2^{CHP}$ |
|---|---|---|---|---|---|---|---|---|---|---|
| 1 | 150 | 135 | 73 | 66.79 | 76.43 | 36.68 | 31.21 | 63.20 | 387.18 | 30.07 |
| 2 | 150 | 135 | 73 | 71.79 | 79.94 | 37.76 | 32.15 | 64.09 | 423.29 | 58.27 |
| 3 | 150 | 135 | 81.09 | 78.95 | 86.12 | 39.83 | 33.88 | 65.68 | 493.20 | 113.57 |
| 4 | 150 | 135 | 98.17 | 84.65 | 91.76 | 41.84 | 35.48 | 67.08 | 559.88 | 166.27 |
| 5 | 150 | 135 | 104.13 | 86.98 | 94.23 | 55.64 | 36.16 | 67.66 | 588.86 | 188.02 |
| 6 | 150 | 135 | 112.49 | 90.50 | 104.35 | 105.64 | 37.22 | 68.53 | 633.29 | 222.88 |
| 7 | 150 | 135 | 113.24 | 90.83 | 154.35 | 123.13 | 37.32 | 68.61 | 637.49 | 226.25 |
| 8 | 150 | 135 | 111.24 | 89.96 | 204.35 | 173.13 | 37.06 | 68.40 | 626.45 | 217.03 |
| 9 | 150 | 135 | 121.49 | 94.57 | 216.43 | 193.64 | 38.46 | 69.51 | 684.67 | 263.33 |
| 10 | 150 | 137.73 | 129.11 | 98.28 | 220.24 | 243.64 | 43.70 | 70.36 | 729.62 | 299.37 |
| 11 | 150 | 143.62 | 135.35 | 101.50 | 223.35 | 186.70 | 93.70 | 71.05 | 766.46 | 328.15 |
| 12 | 150 | 150.95 | 146.63 | 108.02 | 228.96 | 188.67 | 99.16 | 72.25 | 782.18 | 350.67 |
| 13 | 150 | 138.07 | 129.42 | 98.43 | 220.39 | 185.65 | 96.74 | 70.39 | 732.18 | 300.67 |
| 14 | 150 | 135 | 112.67 | 90.58 | 200.43 | 182.77 | 94.20 | 68.55 | 682.18 | 250.67 |

(continued)

**Table 6.9** (continued)

| Hour | $P_1^T$ | $P_2^T$ | $P_3^T$ | $P_4^T$ | $P_5^T$ | $P_6^T$ | $P_7^T$ | $P_8^T$ | $P_1^{CHP}$ | $P_2^{CHP}$ |
|---|---|---|---|---|---|---|---|---|---|---|
| 15 | 150 | 135 | 113.58 | 90.97 | 110.43 | 182.92 | 94.34 | 68.65 | 639.34 | 227.74 |
| 16 | 150 | 135 | 105.67 | 87.61 | 94.91 | 119.59 | 36.35 | 67.81 | 592.29 | 193.57 |
| 17 | 150 | 135 | 91.87 | 82.36 | 89.43 | 169.59 | 34.82 | 66.51 | 542.29 | 144.12 |
| 18 | 150 | 135 | 109.67 | 89.28 | 123.47 | 122.92 | 36.86 | 68.23 | 612.29 | 211.67 |
| 19 | 150 | 135 | 113.34 | 90.87 | 173.47 | 172.92 | 42.99 | 68.62 | 638.03 | 227.53 |
| 20 | 150 | 148.91 | 143.07 | 105.84 | 223.47 | 188.06 | 92.99 | 71.88 | 699.54 | 297.53 |
| 21 | 150 | 153.18 | 150.97 | 110.95 | 211.03 | 120.86 | 99.72 | 72.68 | 649.54 | 247.53 |
| 22 | 150 | 135 | 99.91 | 85.32 | 121.03 | 111.43 | 92.23 | 67.24 | 599.54 | 197.53 |
| 23 | 150 | 135 | 73.09 | 76.93 | 84.26 | 39.18 | 33.35 | 65.21 | 549.54 | 147.53 |
| 24 | 150 | 135 | 73 | 60 | 63.93 | 33.80 | 28.55 | 60.44 | 499.54 | 97.53 |

**Table 6.10** Optimal heat flow and losses for CHPDEED model

| Hour | $H_1^{CHP}$ | $H_2^{CHP}$ | $H_1^H$ | $Loss_k$ |
|---|---|---|---|---|
| 1 | 147.32 | 89 | 153.68 | 13.57 |
| 2 | 150.78 | 89 | 160.22 | 15.28 |
| 3 | 151.95 | 89 | 169.05 | 19.32 |
| 4 | 153.34 | 89 | 177.66 | 24.13 |
| 5 | 163.19 | 89 | 187.81 | 26.68 |
| 6 | 166.08 | 89 | 194.92 | 31.89 |
| 7 | 165.8 | 89 | 195.20 | 34.23 |
| 8 | 169.5 | 89 | 196.50 | 36.61 |
| 9 | 168.51 | 89 | 202.49 | 43.10 |
| 10 | 165.46 | 89 | 205.54 | 50.04 |
| 11 | 168.87 | 89 | 212.13 | 53.87 |
| 12 | 172.81 | 89 | 218.19 | 57.49 |
| 13 | 171.19 | 89 | 209.81 | 49.95 |
| 14 | 169.58 | 89 | 201.42 | 43.06 |
| 15 | 165.67 | 89 | 195.33 | 36.96 |
| 16 | 168.56 | 89 | 192.44 | 28.80 |
| 17 | 155.21 | 89 | 175.79 | 25.98 |
| 18 | 158.26 | 89 | 187.74 | 31.39 |
| 19 | 162.77 | 89 | 193.23 | 36.77 |
| 20 | 159.77 | 89 | 201.23 | 49.28 |
| 21 | 161.08 | 89 | 194.92 | 42.45 |
| 22 | 159.46 | 89 | 186.54 | 31.23 |
| 23 | 143.23 | 89 | 167.77 | 22.08 |
| 24 | 147.46 | 89 | 163.54 | 17.78 |

## 6.5 Incentive-Based Combined Heat and Power Dynamic Economic Emissions Dispatch

DR is considered a means of reshaping the system load profile to achieve improved system efficiency. This concept of DR can be expressed as changes that occur in electricity usage by demand-side consumers from the normal consumption patterns with respect to changes in electricity price over time or financial incentives offered to consumers to reduce electricity usage at times of high wholesale market prices to obtain higher energy efficiency and improve system stability [7]. DR can be classified as price-based DR and incentive-based DR.

Here, we are considering incentive-based CHPDEED (IBCHPDEED). There are two variables: customer power curtailed (× MW) and customer incentive ($ q). Furthermore, we expand the model over more than one scheduling interval and incorporate other practical considerations into the model. The resulting model is detailed thus:

$$\text{Min}\left[\begin{array}{l} u_1\left(\sum_{k=1}^{k}\sum_{j=1}^{J} C_j^T\left(P_{j,k}^T\right)+\sum_{k=1}^{K}\sum_{g=1}^{G} C_g^{\text{CHP}}\left(P_{g,k}^{\text{CHP}},H_{g,k}^{\text{CHP}}\right)+\sum_{k=1}^{K}\sum_{i=1}^{I} C_i^H\left(H_{i,k}^H\right)\right)\\ +u_2\left(\sum_{k=1}^{k}\sum_{j=1}^{J} E_j^T\left(P_{j,k}^T\right)+\sum_{k=1}^{K}\sum_{g=1}^{G} E_g^{\text{CHP}}\left(P_{g,k}^{\text{CHP}}\right)+\sum_{k=1}^{K}\sum_{i=1}^{I} E_i^H\left(H_{i,k}^H\right)\right)\\ +u_3\left(\sum_{k=1}^{K}\sum_{c=1}^{C} q_{c,k}-\alpha_{c,k}x_{c,k}\right)\end{array}\right] \tag{6.35}$$

subject to the following constraints:

$$\sum_{j=1}^{J} P_{j,k}^T+\sum_{g=1}^{G} P_{g,k}^{\text{CHP}}=D_k+\text{Loss}_k \tag{6.36}$$

$$\sum_{g=1}^{G} H_{g,k}^{\text{CHP}}+\sum_{i=1}^{I} H_{i,k}^H=\text{HD}_k+\text{Loss}_k \tag{6.37}$$

$$P_{j,\min}^T\le P_{j,k}^T\le P_{j,\max}^T \tag{6.38}$$

$$P^{\mathrm{CHP}}_{g,\min}\left(H^{\mathrm{CHP}}_{g,k}\right) \le P^{\mathrm{CHP}}_{g,k} \le P^{\mathrm{CHP}}_{g,\max}\left(H^{\mathrm{CHP}}_{g,k}\right) \tag{6.39}$$

$$H^{\mathrm{CHP}}_{g,\min}\left(P^{\mathrm{CHP}}_{g,k}\right) \le H^{\mathrm{CHP}}_{g,k} \le H^{\mathrm{CHP}}_{g,\max}\left(P^{\mathrm{CHP}}_{g,k}\right) \tag{6.40}$$

$$H^{H}_{g,\min} \le H^{H}_{g,k} \le H^{H}_{g,\max} \tag{6.41}$$

$$-\mathrm{DR}^{T}_{j} \le P^{T}_{j,k+1} - P^{T}_{j,k} \le \mathrm{UR}^{T}_{j} \tag{6.42}$$

$$-\mathrm{DR}^{\mathrm{CHP}}_{g} \le P^{\mathrm{CHP}}_{g,t+1} - P^{\mathrm{CHP}}_{g,t} \le \mathrm{UR}^{\mathrm{CHP}}_{g} \tag{6.43}$$

$$\mathrm{Loss}_k = \sum_{j=1}^{J}\sum_{z=1}^{Z} P_{j,k}B_{j,z}P_{z,k}. \tag{6.44}$$

The desired variable result to be obtained by the optimization model is $P^{T}_{j,k}$, $P^{\mathrm{CHP}}_{g,k}$, $H^{\mathrm{CHP}}_{g,k}$, $H^{H}_{i,k}$, $\mathrm{Loss}_k$, $q_{c,k}$, and $x_{c,k}$.

Equation (6.36) is a power balance constraint, which compels the total output power from both thermal- and CHP-generating units to satisfy the load demand and line losses. Constraint (6.37) is a heat balance constraint, which states that the heat output from both heat and CHP units must meet the heat demand. Constraint (6.38) gives the generation limits for thermal-generating units and states that output power from the thermal units must not exceed the allowed limits. The next constraint (6.39) limits the power produced by CHP units within the allowable limits. Constraint (6.40) limits the heat produced by CHP units within allowable the limits. Constraint (6.41) ensures that heat produced by heat-only units is within the allowable limits. Constraints (6.42) and (6.43) are the generator ramp rate limits for the thermal- and CHP-generating units. Equation (6.44) is the transmission line losses constraint and the line losses are determined by the *B*-coefficient method [4].

The data for thermal, CHP, and heat-only generating units are presented in Tables 6.1, 6.2, and 6.3. Similarly, all necessary data are presented in Sect. (6.2). The optimization model is formulated and solved using the CONPOT solver embedded in AIMMS. The AIMMS code for solving the example of an IBCHPDEED problem is shown below:

```
Model Main_Economic_Dispatch_ten_bus_ded {
  Set conventional {
    Index: i, k1;
  }
  Set heat {
    Index: h;
  }
  Set chp {
    Index: p, k2;
  }
  Set Time {
    Index: t;
  }
  Set customer {
    Index: j;
  }
  Parameter pmin {
    IndexDomain: i;
  }
  Parameter pmax {
    IndexDomain: i;
  }
  Parameter a {
    IndexDomain: i;
  }
  Parameter b {
    IndexDomain: i;
  }
  Parameter c {
    IndexDomain: i;
  }
  Parameter e {
    IndexDomain: i;
  }
  Parameter f {
    IndexDomain: i;
  }
  Parameter ma {
    IndexDomain: i;
  }
  Parameter mb {
    IndexDomain: i;
  }
  Parameter mc {
    IndexDomain: i;
  }
  Parameter mem {
    IndexDomain: i;
  }
  Parameter mf {
    IndexDomain: i;
  }
  Parameter w {
    Definition: 0.3;
  }
  Parameter aa {
    IndexDomain: p;
  }
  Parameter bb {
    IndexDomain: p;
  }
  Parameter cc {
    IndexDomain: p;
  }
  Parameter dd {
    IndexDomain: p;
  }
  Parameter ee {
    IndexDomain: p;
```

```
}
Parameter ff {
   IndexDomain: p;
}
Parameter maa {
   IndexDomain: p;
}
Parameter mbb {
   IndexDomain: p;
}
Parameter aaa {
   IndexDomain: h;
}
Parameter ccc {
   IndexDomain: h;
}
Parameter bbb {
   IndexDomain: h;
}
Parameter maaa {
   Definition: 0.0008;
}
Parameter mbbb {
   Definition: 0.0010;
}
Parameter dr {
   IndexDomain: i;
}
Parameter drch {
   IndexDomain: p;
}
Parameter urch {
   IndexDomain: p;
}
Parameter ur {
   IndexDomain: i;
}
Parameter d {
   IndexDomain: t;
}
Parameter he {
   IndexDomain: t;
}
Parameter bma1 {
   IndexDomain: (i,k1);
}
Parameter bma2 {
   IndexDomain: (p,k2);
}
Parameter loss1 {
   IndexDomain: t;
   Definition: sum[i,sum[k1, pc(i,t)*bma1(i,k1)*pc(k1,t)]];
}
Parameter loss2 {
   IndexDomain: t;
   Definition: sum[p,sum[k2, pch(p,t)*bma2(p,k2)*pch(k2,t)]];
}
Parameter w2 {
   Definition: 0.3;
}
Parameter w3 {
   Definition: 0.3;
}
Variable pc {
   IndexDomain: (i,t);
   Range: nonnegative;
}
Variable pch {
   IndexDomain: (p,t);
```

```
    Range: nonnegative;
  }
  Variable hch {
    IndexDomain: (p,t);
    Range: nonnegative;
  }
  Variable hh {
    IndexDomain: (h,t);
    Range: (0, 2695.2);
  }
  Variable y {
    IndexDomain: (j,t)| 9<= ord(t) <=16;
    Range: nonnegative;
  }
  Variable x {
    IndexDomain: (j,t)| 9<= ord(t) <=16;
    Range: nonnegative;
  }
  Variable Total {
    Range: free;
    Definition: {
      (w*(sum[(i,t), a(i)+(b(i)*pc(i,t))+(c(i)*pc(i,t)*pc(i,t))+(e(i)*sin(f(i)*(pmin(i)-pc(i,t))))]+
      sum[(p,t),
aa(p)+(bb(p)*pch(p,t))+(cc(p)*pch(p,t)*pch(p,t))+(dd(p)*hch(p,t))+(ee(p)*hch(p,t)*hch(p,t))+(ff(p)*pch(p,t)*hch(p,t))]+
      sum[(h,t), aaa(h)+(bbb(h)*hh(h,t))+(ccc(h)*hh(h,t)*hh(h,t))]))
      +
      ((w2)*(sum[(i,t), ma(i)+(mb(i)*pc(i,t))+(mc(i)*pc(i,t)*pc(i,t))+(mem(i)*exp(mf(i)*(pc(i,t))))]+
      sum[(p,t), (maa(p)+(mbb(p)))*pch(p,t)]+
      sum[(h,t), ((maaa+mbbb)*hh(h,t))]))
      +
      (w3*(sum[ t, sum[ j, y(j,t)-(lamb(j,t)*x(j,t))]]))
    }
  }
  Parameter totalfuelcost {
    Definition: {
      (sum[(i,t), a(i)+(b(i)*pc(i,t))+(c(i)*pc(i,t)*pc(i,t))+(e(i)*sin(f(i)*(pmin(i)-pc(i,t))))]+
      sum[(p,t),
aa(p)+(bb(p)*pch(p,t))+(cc(p)*pch(p,t)*pch(p,t))+(dd(p)*hch(p,t))+(ee(p)*hch(p,t)*hch(p,t))+(ff(p)*pch(p,t)*hch(p,t))]+
      sum[(h,t), aaa(h)+(bbb(h)*hh(h,t))+(ccc(h)*hh(h,t)*hh(h,t))])
    }
  }
  Parameter totalemissions {
    Definition: (sum[(i,t), ma(i)+(mb(i)*pc(i,t))+(mc(i)*pc(i,t)*pc(i,t))+(mem(i)*exp(mf(i)*(pc(i,t))))]);
  }
  Parameter totalpower {
    Definition: sum[(i,t),pc(i,t)]+sum[(p,t),pch(p,t)];
  }
  Parameter totalheat {
    Definition: sum[(h,t),hh(h,t)]+sum[(p,t),hch(p,t)];
  }
  Parameter totalloss {
    Definition: sum[t,loss1(t)+loss2(t)];
  }
  Parameter totalincentive {
    Definition: sum[(j,t),y(j,t)];
  }
  Parameter totalpowercurtailed {
    Definition: sum[(j,t),x(j,t)];
  }
  Parameter SUMPOWER {
    Definition: sum[(i,t),pc(i,t)]+sum[(p,t),pch(p,t)]+sum[(j,t),x(j,t)];
  }
  Parameter pchmin {
    IndexDomain: p;
  }
  Parameter thet {
    IndexDomain: j;
  }
```

```
Parameter kk1 {
   IndexDomain: j;
}
Parameter kk2 {
   IndexDomain: j;
}
Parameter lamb {
   IndexDomain: (j,t);
}
Parameter tcustcost {
   IndexDomain: j;
   Definition: sum[t,y(j,t)];
}
Parameter totpowcurtailed {
   IndexDomain: j;
   Definition: sum[t,x(j,t)];
}
Parameter CM {
   IndexDomain: j;
}
Parameter hchmin {
   IndexDomain: p;
}
Parameter hchmax {
   IndexDomain: p;
}
Parameter pchmax {
   IndexDomain: p;
}
Parameter hhmin {
   IndexDomain: h;
   Definition: 0;
}
Parameter hhmax {
   IndexDomain: h;
   Definition: 2695.2;
}
Constraint sup {
   IndexDomain: t;
   Definition: sum[i, pc(i,t)]+sum[p, pch(p,t)]=d(t)-sum[j, x(j,t)]+loss1(t)+loss2(t);
}
Constraint supp {
   IndexDomain: t;
   Definition: sum[h, hh(h,t)]+sum[p, hch(p,t)]=he(t);
}
Constraint sup1 {
   IndexDomain: (i,t);
   Definition: pc(i,t)<=pmax(i);
}
Constraint sup2 {
   IndexDomain: (i,t);
   Definition: pc(i,t)>=pmin(i);
}
Constraint sup3 {
   IndexDomain: (i,t) | 1<= ord(t) <=23;
   Definition: pc(i,t)-pc(i,t+1)<=dr(i);
}
Constraint sup4 {
   IndexDomain: (i,t) | 1<= ord(t) <=23;
   Definition: pc(i,t+1)-pc(i,t)<=ur(i);
}
Constraint sup8 {
   IndexDomain: (p,t)| 1<= ord(t) <=23;
   Definition: -drch(p)<=pch(p,t+1)-pch(p,t)<=urch(p);
}
Constraint con1 {
   IndexDomain: (j,t);
   Definition: y(j,t)-((kk1(j)*x(j,t)*x(j,t))+(kk2(j)*x(j,t))-(kk2(j)*x(j,t)*thet(j))) >=0;
}
```

```
Constraint con2 {
  IndexDomain: j;
  Definition:    sum[t,(y(j,t)-((kk1(j)*x(j,t)*x(j,t))+(kk2(j)*x(j,t))-(kk2(j)*x(j,t)*thet(j))))]    >=sum[t,(y(j-1,t)-((kk1(j-1)*x(j-
1,t)*x(j-1,t))+(kk2(j-1)*x(j-1,t))-(kk2(j-1)*x(j-1,t)*thet(j-1))))];
}
Constraint con3 {
  Definition: sum[t,sum[j, y(j,t)]]<=100000;
}
Constraint con4 {
  IndexDomain: j;
  Definition: sum[t,x(j,t)]<=CM(j);
}
Constraint fe1 {
  IndexDomain: (p,t);
  Definition: (pch(p|ord(p)=1,t))+(hch(p|ord(p)=1,t)*(32/180))<=247;
}
Constraint fe2 {
  IndexDomain: (p,t);
  Definition: (pch(p|ord(p)=1,t))-(hch(p|ord(p)=1,t)*(134/75.2))>=-105.745;
}
Constraint fe3 {
  IndexDomain: (p,t);
  Definition: 1=1;
}
Constraint fe4 {
  IndexDomain: (p,t);
  Definition: (pch(p|ord(p)=2,t))+(hch(p|ord(p)=2,t)*(15.6/103.2))<=130.698;
}
Constraint fe5 {
  IndexDomain: (p,t);
  Definition: (pch(p|ord(p)=2,t))-(hch(p|ord(p)=2,t)*(70.2/60.6))>=-46.881;
}
Constraint fe6 {
  IndexDomain: (p,t);
  Definition: (pch(p|ord(p)=2,t))+(hch(p|ord(p)=2,t)*(4/59.1))<=45.076;
}
MathematicalProgram LeastEconDis {
  Objective: Total;
  Direction: minimize;
  Constraints: AllConstraints;
  Variables: AllVariables;
  Type: Automatic;
}
Procedure MainInitialization;
Procedure MainExecution {
  Body: {
    solve LeastEconDis;
  }
}
```

Tables 6.11 and 6.12 give the numerical simulation results obtained from the optimization problem. The results show variation in the hourly power and heat flow from the thermal, CHP and heat-only generators. From the tables, it is obvious that the incorporation of the incentive-based program leads to a reduction in demand across all the load profiles.

**Table 6.11** Optimal power flow for IBCHPDEED model

| Hour | $P_1^T$ | $P_2^T$ | $P_3^T$ | $P_4^T$ | $P_5^T$ | $P_6^T$ | $P_7^T$ | $P_8^T$ | $P_1^{\mathrm{CHP}}$ | $P_2^{\mathrm{CHP}}$ |
|---|---|---|---|---|---|---|---|---|---|---|
| 1 | 150 | 135 | 109.75 | 89.35 | 96.83 | 112.90 | 36.89 | 27.59 | 233.68 | 40 |
| 2 | 150 | 136.45 | 128.18 | 110.10 | 110 | 116.17 | 50 | 29.60 | 234.45 | 40 |
| 3 | 150 | 145.30 | 137.70 | 160.10 | 160 | 118.05 | 80 | 30.71 | 234.35 | 40 |
| 4 | 159.17 | 163.54 | 182.24 | 210.10 | 160 | 130 | 80 | 35.20 | 236.78 | 40 |
| 5 | 160.21 | 164.27 | 251.41 | 215.26 | 160 | 130 | 80 | 35.46 | 234.83 | 40 |
| 6 | 169.17 | 171.44 | 310.50 | 240.97 | 160 | 130 | 80 | 55 | 235.55 | 40 |
| 7 | 172.00 | 174.00 | 317.41 | 290.97 | 160 | 130 | 80 | 55 | 236.21 | 40 |
| 8 | 181.71 | 183.61 | 337.88 | 300 | 160 | 130 | 80 | 55 | 238.20 | 40 |
| 9 | 204.07 | 241.46 | 340 | 300 | 160 | 130 | 80 | 55 | 243.77 | 40 |
| 10 | 206.79 | 321.46 | 340 | 300 | 160 | 130 | 80 | 55 | 244.41 | 40 |
| 11 | 229.00 | 340.70 | 340 | 300 | 160 | 130 | 80 | 55 | 246.90 | 40 |
| 12 | 305.77 | 338.69 | 340 | 300 | 160 | 130 | 80 | 55 | 245.23 | 40 |
| 13 | 225.77 | 332.49 | 340 | 300 | 160 | 130 | 80 | 55 | 246.24 | 40 |
| 14 | 199.45 | 252.49 | 340 | 300 | 160 | 130 | 80 | 55 | 242.59 | 40 |
| 15 | 181.11 | 182.98 | 336.71 | 300 | 160 | 130 | 80 | 55 | 238.57 | 40 |
| 16 | 159.72 | 163.92 | 260.80 | 250 | 160 | 130 | 80 | 35.33 | 233.69 | 40 |
| 17 | 160.55 | 164.51 | 209.71 | 215.52 | 160 | 130 | 80 | 35.55 | 237.01 | 40 |
| 18 | 164.76 | 167.72 | 289.71 | 250 | 160 | 130 | 80 | 55 | 236.20 | 40 |
| 19 | 180.05 | 182.53 | 334.62 | 300 | 160 | 130 | 80 | 55 | 238.81 | 40 |
| 20 | 210.15 | 262.53 | 340.00 | 300 | 160 | 130 | 80 | 55 | 246.22 | 40 |
| 21 | 173.93 | 182.53 | 321.79 | 300 | 160 | 130 | 80 | 55 | 237.21 | 40 |
| 22 | 174.32 | 176.18 | 251.79 | 250 | 160 | 130 | 80 | 55 | 238.37 | 40 |
| 23 | 150 | 144.57 | 171.79 | 200 | 160 | 117.85 | 80 | 30.59 | 235.32 | 40 |
| 24 | 150 | 138.69 | 130.09 | 150 | 110 | 116.53 | 80 | 29.82 | 234.64 | 40 |

**Table 6.12** Optimal heat flow and losses for IBCHPDEED model

| Hour | $H_1^{\mathrm{CHP}}$ | $H_2^{\mathrm{CHP}}$ | $H_1^H$ | $\mathrm{Loss}_k$ |
|---|---|---|---|---|
| 1 | 74.90 | 75.00 | 240.10 | 13.10 |
| 2 | 70.62 | 75.00 | 254.38 | 15.08 |
| 3 | 71.15 | 75.00 | 263.85 | 19.84 |
| 4 | 57.47 | 75.00 | 287.53 | 25.12 |
| 5 | 68.45 | 75.00 | 296.55 | 28.32 |
| 6 | 64.41 | 75.00 | 310.59 | 34.00 |
| 7 | 60.69 | 75.00 | 314.31 | 37.06 |
| 8 | 49.48 | 75.00 | 330.52 | 39.74 |
| 9 | 18.20 | 75.00 | 366.81 | 44.67 |
| 10 | 14.56 | 75.00 | 370.44 | 50.06 |

(continued)

**Table 6.12** (continued)

| Hour | $H_1^{\text{CHP}}$ | $H_2^{\text{CHP}}$ | $H_1^H$ | $\text{Loss}_k$ |
|---|---|---|---|---|
| 11 | 0.56 | 75.00 | 394.44 | 52.90 |
| 12 | 9.93 | 75.00 | 395.07 | 58.09 |
| 13 | 4.25 | 75.00 | 390.75 | 52.09 |
| 14 | 24.82 | 75.00 | 360.18 | 45.05 |
| 15 | 47.42 | 75.00 | 327.58 | 39.62 |
| 16 | 74.89 | 75.00 | 300.11 | 30.18 |
| 17 | 56.18 | 75.00 | 288.82 | 26.58 |
| 18 | 60.73 | 75.00 | 299.27 | 32.95 |
| 19 | 46.06 | 75.00 | 323.94 | 39.42 |
| 20 | 4.37 | 75.00 | 370.63 | 46.41 |
| 21 | 55.05 | 75.00 | 314.95 | 38.34 |
| 22 | 48.52 | 75.00 | 311.48 | 32.08 |
| 23 | 65.73 | 75.00 | 259.27 | 22.45 |
| 24 | 69.52 | 75.00 | 255.48 | 17.37 |

## 6.6 Profit-Based Combined Heat and Power Dynamic Economic Emissions Dispatch

Profit-based DR is widely used in a deregulated market and is intended to maximize the profit for the utility company through selling the energy. In profit-based CHPDEED (PBCHPDEED), the main objective is to maximize profit and minimize emissions. The profit is derived from the difference between the accrued revenue and the operational cost. The optimization problem is given as:

$$\text{Min}\left[ u_1 \left\{ \begin{array}{l} \left( \sum_{k=1}^{k} \sum_{j=1}^{J} C_j^T\left(P_{j,k}^T\right) + \sum_{k=1}^{K} \sum_{g=1}^{G} C_g^{\text{CHP}}\left(P_{g,k}^{\text{CHP}}, H_{g,k}^{\text{CHP}}\right) + \sum_{k=1}^{K} \sum_{i=1}^{I} C_i^H\left(H_{i,k}^H\right) \right) \\ -\left( \sum_{k=1}^{K} \sum_{j=1}^{J} E_k^{\text{price}}\left(P_{j,k}^T\right) + \sum_{k=1}^{K} \sum_{g=1}^{G} E_k^{\text{price}}\left(P_{g,k}^{\text{CHP}}, H_{g,k}^{\text{CHP}}\right) + \sum_{k=1}^{K} \sum_{j=1}^{J} E_k^{\text{price}}\left(H_{i,k}^H\right) \right) \\ + u_2 \left( \sum_{k=1}^{k} \sum_{j=1}^{J} E_j^T\left(P_{j,k}^T\right) + \sum_{k=1}^{K} \sum_{g=1}^{G} E_g^{\text{CHP}}\left(P_{g,k}^{\text{CHP}}\right) + \sum_{k=1}^{K} \sum_{i=1}^{I} E_i^H\left(H_{i,k}^H\right) \right) \end{array} \right\} \right] \tag{6.45}$$

subject to the following constraints:

$$\sum_{j=1}^{J} P_{j,k}^T + \sum_{g=1}^{G} P_{g,k}^{\text{CHP}} \leq D_k + \text{Loss}_k \tag{6.46}$$

$$\sum_{g=1}^{G} H_{g,k}^{\text{CHP}} + \sum_{i=1}^{I} H_{i,k}^{H} \leq \text{HD}_k + \text{Loss}_k \tag{6.47}$$

$$P_{j,\min}^{T} \leq P_{j,k}^{T} \leq P_{j,\max}^{T} \tag{6.48}$$

$$P_{g,\min}^{\text{CHP}}\left(H_{g,k}^{\text{CHP}}\right) \leq P_{g,k}^{\text{CHP}} \leq P_{g,\max}^{\text{CHP}}\left(H_{g,k}^{\text{CHP}}\right) \tag{6.49}$$

$$H_{g,\min}^{\text{CHP}}\left(P_{g,k}^{\text{CHP}}\right) \leq H_{g,k}^{\text{CHP}} \leq H_{g,\max}^{\text{CHP}}\left(P_{g,k}^{\text{CHP}}\right) \tag{6.50}$$

$$H_{g,\min}^{H} \leq H_{g,k}^{H} \leq H_{g,\max}^{H} \tag{6.51}$$

$$-\text{DR}_j^{T} \leq P_{j,k+1}^{T} - P_{j,k}^{T} \leq \text{UR}_j^{T} \tag{6.52}$$

$$-\text{DR}_g^{\text{CHP}} \leq P_{g,t+1}^{\text{CHP}} - P_{g,t}^{\text{CHP}} \leq \text{UR}_g^{\text{CHP}} \tag{6.53}$$

$$\text{Loss}_k = \sum_{j=1}^{J} \sum_{z=1}^{Z} P_{j,k} B_{j,z} P_{z,k} \tag{6.54}$$

The desired variable result to be obtained by the optimization model is $P_{j,k}^{T}$, $P_{g,k}^{\text{CHP}}$, $H_{g,k}^{\text{CHP}}$, $H_{i,k}^{H}$ and $\text{Loss}_k$.

Equation (6.46) is a power balance constraint, which states that the power produced by the generators can be less than the total load demand and line losses. Similarly, constraint (6.47) is a heat balance constraint, which states that the heat output from both heat and CHP units must be less than the heat demand. Constraint (6.48) gives the generation limits for thermal-generating units and states that output power from the thermal units must not exceed the allowed limits. The next constraint (6.49) limits the power produced by CHP units within the allowable limits. Constraint (6.50) limits the heat produced by CHP units within the allowable limits. Constraint (6.51) ensures that heat produced by heat-only units is within the allowable limits. Constraints (6.52) and (6.53) are the generator ramp rate limits for the thermal- and CHP-generating units. Equation (6.54) is the transmission line losses constraint and the line losses are determined by the B-coefficient method [4].

The data for thermal, CHP, and heat-only generating units are presented in Tables 6.1, 6.2, and 6.3. Similarly, all necessary data are presented in Sect. (6.2). The optimization model is formulated and solved using the CONPOT solver embedded in AIMMS. The AIMMS code for solving an example of a PBCHPDEED problem is shown below:

```
Model Main_Economic_Dispatch_ten_bus_ded {
  Set conventional {
    Index: i, k1;
  }
  Set heat {
    Index: h;
  }
  Set chp {
    Index: p, k2;
  }
  Set Time {
    Index: t;
  }
  Set customer {
    Index: j;
  }
  Parameter pmin {
    IndexDomain: i;
  }
  Parameter pmax {
    IndexDomain: i;
  }
  Parameter a {
    IndexDomain: i;
  }
  Parameter b {
    IndexDomain: i;
  }
  Parameter c {
    IndexDomain: i;
  }
  Parameter e {
    IndexDomain: i;
  }
  Parameter f {
    IndexDomain: i;
  }
  Parameter ma {
    IndexDomain: i;
  }
  Parameter mb {
    IndexDomain: i;
  }
  Parameter mc {
    IndexDomain: i;
  }
  Parameter mem {
    IndexDomain: i;
  }
  Parameter mf {
    IndexDomain: i;
  }
  Parameter w {
    Definition: 0.3;
  }
  Parameter aa {
    IndexDomain: p;
  }
  Parameter bb {
    IndexDomain: p;
  }
  Parameter cc {
```

```
    IndexDomain: p;
}
Parameter dd {
    IndexDomain: p;
}
Parameter ee {
    IndexDomain: p;
}
Parameter ff {
    IndexDomain: p;
}
Parameter maa {
    IndexDomain: p;
}
Parameter mbb {
    IndexDomain: p;
}
Parameter aaa {
    IndexDomain: h;
}
Parameter ccc {
    IndexDomain: h;
}
Parameter bbb {
    IndexDomain: h;
}
Parameter maaa {
    Definition: 0.0008;
}
Parameter mbbb {
    Definition: 0.0010;
}
Parameter dr {
    IndexDomain: i;
}
Parameter drch {
    IndexDomain: p;
}
Parameter urch {
    IndexDomain: p;
}
Parameter ur {
    IndexDomain: i;
}
Parameter d {
    IndexDomain: t;
}
Parameter he {
    IndexDomain: t;
}
Parameter bma1 {
    IndexDomain: (i,k1);
}
Parameter bma2 {
    IndexDomain: (p,k2);
}
Parameter loss1 {
    IndexDomain: t;
    Definition: sum[i,sum[k1, pc(i,t)*bma1(i,k1)*pc(k1,t)]];
}
Parameter loss2 {
    IndexDomain: t;
```

```
    Definition: sum[p,sum[k2, pch(p,t)*bma2(p,k2)*pch(k2,t)]];
  }
  Parameter w2 {
    Definition: 0.3;
  }
  Parameter w3 {
    Definition: 0.3;
  }
  Variable pc {
    IndexDomain: (i,t);
    Range: nonnegative;
  }
  Variable pch {
    IndexDomain: (p,t);
    Range: nonnegative;
  }
  Variable hch {
    IndexDomain: (p,t);
    Range: nonnegative;
  }
  Variable hh {
    IndexDomain: (h,t);
    Range: (0, 2695.2);
  }
  Variable y {
    IndexDomain: (j,t)| 9<= ord(t) <=16;
    Range: nonnegative;
  }
  Variable x {
    IndexDomain: (j,t)| 9<= ord(t) <=16;
    Range: nonnegative;
  }
  Variable Total {
    Range: free;
    Definition: {
      (w*(sum[(i,t), a(i)+(b(i)*pc(i,t))+(c(i)*pc(i,t)*pc(i,t))+(e(i)*sin(f(i)*(pmin(i)-pc(i,t))))]+
      sum[(p,t),
aa(p)+(bb(p)*pch(p,t))+(cc(p)*pch(p,t)*pch(p,t))+(dd(p)*hch(p,t))+(ee(p)*hch(p,t)*hch(p,t))+(ff(p)*pch(p,t)*hch(p,
t))]+
      sum[(h,t), aaa(h)+(bbb(h)*hh(h,t))+(ccc(h)*hh(h,t)*hh(h,t))]))
      +
      ((w2)*(sum[(i,t), ma(i)+(mb(i)*pc(i,t))+(mc(i)*pc(i,t)*pc(i,t))+(mem(i)*exp(mf(i)*(pc(i,t))))]+
      sum[(p,t), (maa(p)+(mbb(p)))*pch(p,t)]+
      sum[(h,t), ((maaa+mbbb)*hh(h,t))]))
      +
      (w3*(sum[ t, sum[ j, y(j,t)-(lamb(j,t)*x(j,t))]]))
    }
  }
  Parameter totalfuelcost {
    Definition: {
      (sum[(i,t), a(i)+(b(i)*pc(i,t))+(c(i)*pc(i,t)*pc(i,t))+(e(i)*sin(f(i)*(pmin(i)-pc(i,t))))]+
      sum[(p,t),
aa(p)+(bb(p)*pch(p,t))+(cc(p)*pch(p,t)*pch(p,t))+(dd(p)*hch(p,t))+(ee(p)*hch(p,t)*hch(p,t))+(ff(p)*pch(p,t)*hch(p,
t))]+
      sum[(h,t), aaa(h)+(bbb(h)*hh(h,t))+(ccc(h)*hh(h,t)*hh(h,t))])
    }
  }
  Parameter totalemissions {
    Definition: (sum[(i,t), ma(i)+(mb(i)*pc(i,t))+(mc(i)*pc(i,t)*pc(i,t))+(mem(i)*exp(mf(i)*(pc(i,t))))]);
  }
  Parameter totalpower {
```

```
  Definition: sum[(i,t),pc(i,t)]+sum[(p,t),pch(p,t)];
}
Parameter totalheat {
  Definition: sum[(h,t),hh(h,t)]+sum[(p,t),hch(p,t)];
}
Parameter totalloss {
  Definition: sum[t,loss1(t)+loss2(t)];
}
Parameter totalincentive {
  Definition: sum[(j,t),y(j,t)];
}
Parameter totalpowercurtailed {
  Definition: sum[(j,t),x(j,t)];
}
Parameter SUMPOWER {
  Definition: sum[(i,t),pc(i,t)]+sum[(p,t),pch(p,t)]+sum[(j,t),x(j,t)];
}
Parameter pchmin {
  IndexDomain: p;
}
Parameter thet {
  IndexDomain: j;
}
Parameter kk1 {
  IndexDomain: j;
}
Parameter kk2 {
  IndexDomain: j;
}
Parameter lamb {
  IndexDomain: (j,t);
}
Parameter tcustcost {
  IndexDomain: j;
  Definition: sum[t,y(j,t)];
}
Parameter totpowcurtailed {
  IndexDomain: j;
  Definition: sum[t,x(j,t)];
}
Parameter CM {
  IndexDomain: j;
}
Parameter hchmin {
  IndexDomain: p;
}
Parameter hchmax {
  IndexDomain: p;
}
Parameter pchmax {
  IndexDomain: p;
}
Parameter hhmin {
  IndexDomain: h;
  Definition: 0;
}
Parameter hhmax {
  IndexDomain: h;
  Definition: 2695.2;
}
Constraint sup {
  IndexDomain: t;
```

```
    Definition: sum[i, pc(i,t)]+sum[p, pch(p,t)]=d(t)-sum[j, x(j,t)]+loss1(t)+loss2(t);
  }
  Constraint supp {
    IndexDomain: t;
    Definition: sum[h, hh(h,t)]+sum[p, hch(p,t)]=he(t);
  }
  Constraint sup1 {
    IndexDomain: (i,t);
    Definition: pc(i,t)<=pmax(i);
  }
  Constraint sup2 {
    IndexDomain: (i,t);
    Definition: pc(i,t)>=pmin(i);
  }
  Constraint sup3 {
    IndexDomain: (i,t) | 1<= ord(t) <=23;
    Definition: pc(i,t)-pc(i,t+1)<=dr(i);
  }
  Constraint sup4 {
    IndexDomain: (i,t) | 1<= ord(t) <=23;
    Definition: pc(i,t+1)-pc(i,t)<=ur(i);
  }
  Constraint sup8 {
    IndexDomain: (p,t)| 1<= ord(t) <=23;
    Definition: -drch(p)<=pch(p,t+1)-pch(p,t)<=urch(p);
  }
  Constraint con1 {
    IndexDomain: (j,t);
    Definition: y(j,t)-((kk1(j)*x(j,t)*x(j,t))+(kk2(j)*x(j,t))-(kk2(j)*x(j,t)*thet(j))) >=0;
  }
  Constraint con2 {
    IndexDomain: j;
    Definition: sum[t,(y(j,t)-((kk1(j)*x(j,t)*x(j,t))+(kk2(j)*x(j,t))-(kk2(j)*x(j,t)*thet(j))))] >=sum[t,(y(j-1,t)-((kk1(j-
1)*x(j-1,t)*x(j-1,t))+(kk2(j-1)*x(j-1,t))-(kk2(j-1)*x(j-1,t)*thet(j-1))))];
  }
  Constraint con3 {
    Definition: sum[t,sum[j, y(j,t)]]<=100000;
  }
  Constraint con4 {
    IndexDomain: j;
    Definition: sum[t,x(j,t)]<=CM(j);
  }
  Constraint fe1 {
    IndexDomain: (p,t);
    Definition: (pch(p|ord(p)=1,t))+(hch(p|ord(p)=1,t)*(32/180))<=247;
  }
  Constraint fe2 {
    IndexDomain: (p,t);
    Definition: (pch(p|ord(p)=1,t))-(hch(p|ord(p)=1,t)*(134/75.2))>=-105.745;
  }
  Constraint fe3 {
    IndexDomain: (p,t);
    Definition: 1=1;
  }
  Constraint fe4 {
    IndexDomain: (p,t);
    Definition: (pch(p|ord(p)=2,t))+(hch(p|ord(p)=2,t)*(15.6/103.2))<=130.698;
  }
  Constraint fe5 {
    IndexDomain: (p,t);
    Definition: (pch(p|ord(p)=2,t))-(hch(p|ord(p)=2,t)*(70.2/60.6))>=-46.881;
  }
```

```
Constraint fe6 {
   IndexDomain: (p,t);
   Definition: (pch(p|ord(p)=2,t))+(hch(p|ord(p)=2,t)*(4/59.1))<=45.076;
}
MathematicalProgram LeastEconDis {
   Objective: Total;
   Direction: minimize;
   Constraints: AllConstraints;
   Variables: AllVariables;
   Type: Automatic;
}
Procedure MainInitialization;
Procedure MainExecution {
   Body: {
      solve LeastEconDis;
   }
}
```

Tables 6.13 and 6.14 present the numerical simulation results obtained from the optimization problem from the PBCHPDEED model. The results show variation in the hourly power and heat flow from the thermal, CHP, and heat-only generators. It can be seen from the results that integration of a profit-based DR program leads to a reduction in the cost of power generation.

**Table 6.13** Optimal power flow for PBCHPDEED model

| Hour | $P_1^T$ | $P_2^T$ | $P_3^T$ | $P_4^T$ | $P_5^T$ | $P_6^T$ | $P_7^T$ | $P_8^T$ | $P_1^{CHP}$ | $P_2^{CHP}$ |
|---|---|---|---|---|---|---|---|---|---|---|
| 1 | 150 | 135 | 116.90 | 92.51 | 100.49 | 114.10 | 37.84 | 28.34 | 234.37 | 40 |
| 2 | 150 | 141.01 | 132.42 | 119.58 | 110.72 | 116.99 | 50 | 30.09 | 234.88 | 40 |
| 3 | 150 | 149.35 | 143.93 | 169.58 | 160 | 119.355 | 80.00 | 234.97 | 40.00 | 67.67 |
| 4 | 160.85 | 164.72 | 189.19 | 215.76 | 160 | 130 | 80 | 55 | 237.06 | 40 |
| 5 | 159.76 | 163.95 | 259.57 | 226.97 | 160 | 130 | 80 | 55 | 234.75 | 40 |
| 6 | 177.46 | 179.24 | 329.36 | 276.97 | 160 | 130 | 80 | 55 | 237.59 | 40 |
| 7 | 195.76 | 200.29 | 340 | 300 | 160 | 130 | 80 | 55 | 242.66 | 40 |
| 8 | 195.27 | 280.29 | 340 | 300 | 160 | 130 | 80 | 55 | 241.99 | 40 |
| 9 | 193.99 | 204.36 | 340 | 300 | 160 | 130 | 80 | 55 | 241.11 | 40 |
| 10 | 198.95 | 284.36 | 340 | 300 | 160 | 130 | 80 | 55 | 242.46 | 40 |
| 11 | 212.42 | 332.49 | 340 | 300 | 160 | 130 | 80 | 55 | 244.57 | 40 |
| 12 | 241.57 | 342.08 | 340 | 300 | 160 | 130 | 80 | 55 | 246.26 | 40 |
| 13 | 215.12 | 300.08 | 340 | 300 | 160 | 130 | 80 | 55 | 245.08 | 40 |
| 14 | 187.55 | 220.08 | 340 | 300 | 160 | 130 | 80 | 55 | 239.30 | 40 |
| 15 | 214.48 | 255.78 | 340 | 300 | 160 | 130 | 80 | 55 | 247.00 | 40 |
| 16 | 164.70 | 175.78 | 297.65 | 250 | 160 | 130 | 80 | 55 | 234.60 | 40 |
| 17 | 159.72 | 163.92 | 261.08 | 223.37 | 160 | 130 | 80 | 55 | 236.87 | 40 |
| 18 | 176.63 | 183.98 | 327.64 | 273.37 | 160 | 130 | 80 | 55 | 238.97 | 40 |
| 19 | 207.23 | 263.98 | 340 | 300 | 160 | 130 | 80 | 55 | 246.11 | 40 |
| 20 | 205.67 | 215.03 | 340 | 300 | 160 | 130 | 80 | 55 | 245.00 | 40 |
| 21 | 170.98 | 173.06 | 314.98 | 300 | 160 | 130 | 80 | 55 | 236.50 | 40 |
| 22 | 190.47 | 193.37 | 299.48 | 274.39 | 160 | 130 | 80 | 55 | 242.78 | 40 |
| 23 | 150 | 144.10 | 219.48 | 224.39 | 114.17 | 117.73 | 80 | 30.52 | 235.25 | 40 |
| 24 | 150 | 135 | 139.48 | 180.51 | 101.01 | 114.27 | 80 | 28.45 | 233.40 | 40 |

**Table 6.14** Optimal heat flow and losses for PBCHPDEED model

| Hour | $H_1^{\text{CHP}}$ | $H_2^{\text{CHP}}$ | $H_1^H$ | $\text{Loss}_k$ |
|---|---|---|---|---|
| 1 | 71.04 | 75.00 | 243.96 | 13.55 |
| 2 | 68.18 | 75.00 | 256.82 | 15.70 |
| 3 | 75.00 | 267.33 | 210.62 | 20.25 |
| 4 | 55.89 | 75.00 | 289.11 | 26.58 |
| 5 | 68.88 | 75.00 | 296.12 | 30.00 |
| 6 | 52.95 | 75.00 | 322.06 | 37.61 |
| 7 | 24.43 | 75.00 | 350.57 | 41.70 |
| 8 | 28.17 | 75.00 | 351.83 | 46.55 |
| 9 | 33.14 | 75.00 | 351.86 | 41.80 |
| 10 | 25.56 | 75.00 | 359.44 | 47.05 |
| 11 | 13.69 | 75.00 | 381.31 | 51.19 |
| 12 | 4.16 | 75.00 | 400.84 | 53.82 |
| 13 | 10.80 | 75.00 | 384.20 | 49.15 |
| 14 | 43.30 | 75.00 | 341.70 | 42.31 |
| 15 | 123.00 | 75.00 | 375.00 | 46.27 |
| 16 | 69.80 | 75.00 | 305.23 | 33.73 |
| 17 | 56.97 | 75.00 | 288.00 | 29.95 |
| 18 | 45.19 | 75.00 | 315.00 | 37.58 |
| 19 | 5.01 | 75.00 | 364.99 | 46.32 |
| 20 | 10.06 | 75.00 | 364.94 | 43.20 |
| 21 | 59.07 | 75.00 | 310.93 | 37.29 |
| 22 | 23.74 | 75.00 | 336.26 | 37.48 |
| 23 | 66.08 | 75.00 | 258.92 | 23.64 |
| 24 | 76.49 | 75.00 | 248.51 | 18.12 |

## 6.7 Summary

This chapter presented the incorporation of incentive-based and profit-based DR programs with the CHPDEED. From the modeling and simulation, it is established that CHPDEED with DR programs are complicated formulations ensuring optimality at both the supply and demand side of the power system.

## 6.8 Nomenclature

The nomenclature used in this chapter is given below:

**Set**

$c$ customer
$i$ heat unit
$j$ thermal unit

$k$ time
$g$ CHP unit

**Parameters**

$C_j^T$ fuel cost for thermal-generating unit $j$
$C_g^{\text{CHP}}$ fuel cost for combined heat and power-generating unit $g$
$C_i^H$ fuel cost for heat-generating unit $i$
$D_k$ total system power demand at time $k$
$\text{Loss}_k$ system losses
$P_{j,\min}^T$ minimum power capacity of thermal-generating unit $j$
$P_{j,\max}^T$ maximum power capacity of thermal-generating unit $j$
$P_{g,\min}^{\text{CHP}}$ minimum power capacity of combined heat- and power-generating unit $g$
$P_{g,\max}^{\text{CHP}}$ maximum power capacity of combined heat- and power-generating unit $g$
$H_{g,\min}^H$ minimum heat capacity of combined heat- and power-generating unit $g$
$H_{g,\max}^H$ maximum heat capacity of combined heat- and power-generating unit $g$
$\text{DR}_j^T$ maximum ramp-down of thermal-generating unit $j$
$\text{UR}_j^T$ maximum ramp-up rates of thermal-generating unit $j$
$\text{DR}_g^{\text{CHP}}$ maximum ramp-down rates of combined heat- and power-generating unit $g$
$\text{UR}_g^{\text{CHP}}$ maximum ramp-up rates of combined heat- and power-generating unit $g$
$E_i^H$ emissions cost for heat-generating unit $i$
$E_j^T$ emissions cost for thermal-generating unit $j$
$E_g^{\text{CHP}}$ emissions cost for combined heat- and power-generating unit $g$
$\text{HD}_k$ total system heat demand at time $k$
$B_{j,z}$ element of the loss coefficient square matrix
$\alpha_{c,k}$ power interruptible value for different customer $c$ at time $k$
$E_k^{\text{price}}$ predicted price of energy at time $k$

**Variables**

$x_{c,k}$ quantity of curtailed power by customer $c$ at time $k$
$q_{c,k}$ amount of incentive paid to customer $c$ at time $k$
$P_{j,k}^T$ power produced by thermal-generating unit $j$ at time $k$
$P_{g,k}^{\text{CHP}}$ power produced by combined heat- and power-generating unit g at time k
$H_{g,k}^{\text{CHP}}$ heat produced by combined heat- and power-generating unit $g$ at time $k$
$H_{i,k}^H$ heat produced by heat-generating unit $i$ at time $k$

## References

1. Elaiw AM, Xia X, Shehata AM (2013) Combined heat and power dynamic economic dispatch with emission limitations using hybrid DE-SQP method. Hindawi Publ. Corp., pp 13–15
2. Gbadamosi SL, Nwulu NI, Sun Y (2018a) Harmonic and power loss minimization in power systems incorporating renewable energy sources and locational marginal pricing. J Renew Sustain Energy 10:1–24
3. Gbadamosi SL, Nwulu NI (2020) Reliability assessment of composite generation and transmission expansion planning incorporating renewable energy sources. J Renew Sustain Energy 026301(12)
4. Nwulu NI, Xia X (2015a) Implementing a model predictive control strategy on the dynamic economic emission dispatch problem with game theory based demand response programs. Energy 91:404–419
5. Nwulu NI, Xia X (2017) Optimal dispatch for a microgrid incorporating renewables and demand response. Renew Energy 101:16–28
6. Gbadamosi SL, Nwulu NI, Sun Y (2018b) Multi-objective optimisation for composite generation and transmission expansion planning considering offshore wind power and feed-in tariffs. IET Renew Power Gener 12(14):1687–1697
7. Nwulu NI, Xia X (2015b) Multi-objective dynamic economic emission dispatch of electric power generation integrated with game theory based demand response programs. Energy Convers Manag 89:963–974

# Chapter 7
# Hybrid Grid-Connected Renewable Energy Sources Powered Microgrid with Demand Response

## 7.1 Introduction

The microgrid is referred to as a low-voltage power grid consisting of distributed generators and storage systems, as well as controllable and critical loads located near the users. Microgrids can be powered through conventional generators or RES. The RES can be either solar or wind power, while the conventional generating units include diesel or thermal generators. The storage systems widely used in microgrids include flywheel, pump hydro, battery, super capacitor, and hybrid storage systems. In modern power systems, a microgrid can be operated in grid-connected or off-grid mode. This chapter provides a comprehensive framework for microgrid energy management problems with incentive-based DR. In this study, the users are allowed to view the economic benefit associated with incentive-based DR and they are rewarded for their degree of participation.

## 7.2 Hybrid Renewable Energy Sources

Hybrid RES (HRES) form a system that supplies power to electrical loads from two or more RES [1]. The electrical loads can be either an AC or DC load. The sources of energy can be a conventional power system, renewable resources, or energy storage systems. The RES consists of solar PV and wind power sources. A hybrid renewable system is essential to ensure good reliability and higher economic feasibility for both grid-connected and off-grid systems [2]. This approach ensures that there is no lack of energy supply, as some energy sources serve as complementary for strengthening the system.



N. Nwulu and S. L. Gbadamosi, *Optimal Operation and Control of Power Systems Using an Algebraic Modelling Language*, Green Energy and Technology,
https://doi.org/10.1007/978-3-030-00395-1_7

### 7.2.1 Wind Power

Electricity generation in a wind turbine is achieved through extraction of the available kinetic energy from wind, which is converted into electrical energy [3, 4]. This is done through a rotor consisting of two or more rotating blades interconnected with a generator. The amount of wind energy received depends on the location and availability of wind speed. The wind energy produced can be estimated as [5]:

$$P_w = \frac{1}{2}\rho V^3 A\eta \tag{7.1}$$

where $\rho$ represents the air density, $V$ is the wind speed, $A$ is the swept area of the turbine, and $\eta$ is the maximum power extraction efficiency of the wind generator.

### 7.2.2 Solar Photovoltaic

Electricity is produced by solar radiation through conversion of sunlight into electrical energy through harvesting of light photons to obtain electrons. The power output and efficiency of a PV system depends on the days and seasons of the year owing to global solar radiation. The PV power output is given by the following equations [6, 7]:

$$P_{\text{pv}} = V * I * n_m * \text{FF} \tag{7.2}$$

$$I = s(t) * (I^{\text{sc}} * K * (T_{\text{ct}} - 25)) \tag{7.3}$$

$$V = V^{\text{oc}} + K_v * T_{\text{ct}} \tag{7.4}$$

$$T_{\text{ct}} = T_{\text{at}} + s(t) * \left(\frac{\text{NOCT} - 25}{0.8}\right) \tag{7.5}$$

$$\text{FF} = \frac{V^{\max} * I^{\max}}{V^{\text{oc}} * I^{\text{sc}}} \tag{7.6}$$

where $P_{\text{pv}}$ represents the power output of the PV in watts (W), $V$ is the output voltage (V), $I$ is terminal current (A), FF is the fill factor, $V^{\max}$ and $I^{\max}$ are the voltage (V) and current (A) at peak power, respectively, $I^{\text{sc}}$ is the short circuit current (A), $V^{\text{oc}}$ is the open circuit voltage (V), $K$ is the current temperature coefficient (mA/°C), $T_{\text{ct}}$ is the cell temperature (°C), $K_v$ is the coefficient of voltage temperature (mV/°C), $T_{\text{at}}$ is the ambient temperature (°C), NOCT is the nominal cell temperature (°C), and $s(t)$ is the random irradiance.

## 7.3 Microgrid Storage Systems

In a microgrid, a storage system is necessary to avoid significant demand variations and to ensure a steady supply of energy. There are many benefits associated with energy storage, which include supply and demand stability, improving power quality, regulation of grid voltage and frequency and normalizing intermittency associated with RES [8, 9]. A storage system coordinates the integration of RES into a microgrid through smoothing of the fluctuation and reactive power compensation of the RES. Various technologies are used for an energy storage system to store electrical energy in a microgrid and these are discussed below.

### 7.3.1 *Battery Storage System*

Battery storage systems have been widely used as solution to the intermittencies and fluctuations of RES because of their environmental independence and adequate response capability [10]. These are mostly available in hybrid storage systems and ensure adequate balance between electricity supply and demand. Many factors contribute to battery sizing, such as the storage capacity, temperature, lifetime, charging, and discharging of the battery. Therefore, the following equations depict the charging and discharging of the battery [11]:

$$E_b^t = E_b^{t-1}(1-\partial) + \eta_b\left[E_i^t - \frac{E_{\text{load}}^t}{\eta_{\text{inv}}}\right] \tag{7.7}$$

where $E_b^t$ and $E_b^{t-1}$ are the batteries' stored energy at a time ($t$) and previous time ($t-1$), respectively, $E_i^t$ is the power produced by the generating units $i$, $E_{\text{load}}^t$ is the load demand in time ($t$), $\eta_{\text{inv}}$ and $\eta_b$ are inverter and battery efficiencies, respectively. The battery storage of the system will be discharged when the load demands are more than the energy produced by the generators and is given as:

$$E_b^t = E_b^t(1-\partial) - \left[\frac{E_{\text{load}}^t}{\eta_{\text{inv}}} - E_i^t\right] \tag{7.8}$$

### 7.3.2 *Pumped Storage Hydro System*

An energy storage system plays a vital role in balancing RES' contributions to power generation. This is possible owing to the intermittent nature of RES and load variability. A pumped storage hydro (PSH) system is often seen as a mature and efficient storage system capable of accommodating the variability nature of RES in microgrid systems [12]. A PSH system consists of a pump, a hydro turbine, and two reservoirs (upper and lower); it stores electricity in the form of gravitational potential energy [13]. The water is pumped from the lower reservoir to the upper reservoir, when the power produced is lower than the demand, but the water stored is released back through the hydro turbine into the lower reservoir to produce electricity. Therefore, the design of a PSH system depends on the size of reservoirs, types of pumps and turbines. Equation (7.9) describes the power used to pump water from the lower reservoir to the upper reservoir.

$$P_p = P_{\text{input}}\eta_m \tag{7.9}$$

where $P_{\text{input}}$ represents the electrical input power to the motor, $\eta_m$ is the efficiency of the motor.

The flow rate of water is given as:

$$Q_p = \frac{P_p\eta_p}{Hg\rho} \tag{7.10}$$

where $\eta_p$ is the pump efficiency, $g$ is the gravitational acceleration, $\rho$ is the density of water, and $H$ is the difference in pump head between the two reservoirs.

The power output of the hydro turbine is given as:

$$P_t = H_t Q_t \eta_t \rho g \tag{7.11}$$

where $H_t$ represents the turbine head, $Q_t$ is the flow rate of the turbine, and $\eta_t$ is the turbine efficiency.

### 7.3.3 Flywheel Storage System

The flywheel storage system is one of the oldest and most common storage systems. The flywheel storage system is an electromechanical system, which can save energy in the form of kinetic energy and regenerates the stored energy when needed [14]. The quantity of energy stored is based on the rotational speed and mass of the flywheel. It comprises a disk, an electrical machine, DC-link capacitor and a converter. Its ability to charge and discharge easily in a microgrid and its environmentally friendly nature makes it a good choice for renewable energy integration. The energy stored in the flywheel storage system can be estimated as [15]:

$$E = \frac{1}{2}Jw^2 \tag{7.12}$$

where $E$ represents the energy stored in the flywheel, $w$ is the angular velocity of the rotor and $J$ is the moment of inertia of the rotor. This moment of inertia is a function of the mass and shape of the flywheel and is given as:

$$J = \frac{1}{2} m r^2 \tag{7.13}$$

where $m$ and $r$ are the mass and radius of the solid cylindrical rotor, respectively.

### *7.3.4 Supercapacitor Storage System*

A supercapacitor storage system is one of the most effective devices for optimal operation of a microgrid owing to its long lifespan, no internal chemical changes, no demand for maintenance, and no equipment with moving parts. It has a long lifecycle and the ability to charge and discharge several times with low energy density [16]. The supercapacitor storage system comprises the capacitor array for storing energy, an inverter that converts the DC power stored into AC power, and a microprocessor for integrating the supercapacitor to the power system. Similar to a normal capacitor, the stored energy in a supercapacitor is given as [17]:

$$E = \frac{1}{2} C V^2 \tag{7.14}$$

where $C$ and $V$ are the capacitance and voltage across the capacitor, respectively.

### *7.3.5 Hybrid Storage System*

Individual energy storage systems have proven not to be technically and economically viable owing to their special attributes that render them more suitable for different applications. These attributes include power density, cost of operation, response time, and energy density [3, 18]. The combination of two or more energy storage systems increases power system security and reliability in a microgrid with renewable energy. In the literature, several configurations of hybrid energy storage systems have been considered, such as battery and compressed air, flywheel and battery, supercapacitor and battery, compressed air and supercapacitor, and lithium-ion and battery systems, and so on [19]. The formation of a hybrid system from different storage systems could be based on the selection that gives a better net present value or technical and economic constraints.

## 7.4 Hybrid Renewable Energy Sources with Battery Storage in a Microgrid

HRES systems comprise a combination of two or more renewable sources used together to reduce the variable nature of the RES, thus improving the energy security, efficiency, and energy balance [11]. In microgrids, RES has become the systems of choice owing to their environmentally friendly nature and cost efficiency over other sources [20, 21]. The objective is to minimize the operational cost and determine the optimal daily power dispatch of the microgrid. Therefore, this chapter analyzes the economic benefit of powering the microgrid with RES and a battery system.

The proposed optimization model seeks to minimize the daily operational costs by considering RES and a battery system. The problem formulation is expressed as:

$$\text{Min}\left[\sum_{t=1}^{T}\sum_{k=1}^{I}\left(a_k + b_k P_{k,t} + c_k (P_{k,t})^2\right)\right] \tag{7.15}$$

subject to the following constraints:

$$\sum_{i} P_{i,t} + P_{w,t} + P_{v,t} = d_t \tag{7.16}$$

$$P_k^{\min} \le P_{k,t} \le P_k^{\max} \tag{7.17}$$

$$0 \le P_{s,t} \le S_t \tag{7.18}$$

$$0 \le P_{w,t} \le W_t \tag{7.19}$$

$$P_{v,t} + P_{B,t} \le P_{s,t} \tag{7.20}$$

$$B_{\text{soc}}(t) = B_{\text{soc}(0)} - \sigma \sum_{\gamma=1}^{t} P_B(\gamma) \quad 1 \le \gamma \le t \tag{7.21}$$

$$B_{\text{soc}}^{\min} \le B_{\text{soc}(0)} - \partial \sum_{\gamma=1}^{n} P_B(\gamma) \le B_{\text{soc}}^{\max} \tag{7.22}$$

$$B_{\text{soc}}^{\min} = (1 - \text{DoD}) B_{\text{soc}}^{\max} \tag{7.23}$$

The desired variable result to be obtained by the optimization model is $P_{k,t}$, $P_{v,t}$, $P_{w,t}$, and $P_{B,t}$.

Equation (7.16) is a power balance constraint ensuring that at any time t, the power produced by conventional generators, wind, and PV generating units equals the total loads. Constraints (7.17)–(7.19) are the generating limits for conventional, solar, and wind generating units, respectively. Constraint (7.20) states that the sum of charging and discharging of battery power and power fed into the load must be less or equal to the total forecast PV. Constraint (7.21) describes the charging and discharging of the battery, which is a function of PV and battery power output at any given time. Constraints (7.22) and (7.23) are the battery state of charge limit; it states that the available battery capacity must be within the allowable capacity.

In this study, the data for the conventional generator is presented in Table 7.1. The technical specifications for the proposed HRES and the load demand are presented in Tables 7.2 and 7.3.

**Table 7.1** Data for conventional generator

| Generator | $a_i$ | $b_i$ | $c_i$ | $P_{i,\ min}$ | $P_{i,\ max}$ |
|---|---|---|---|---|---|
| 1 | 2650 | 14.5 | 0.0345 | 0 | 50 |

**Table 7.2** Technical specifications for the RES

| Components | Capacity (kW) |
|---|---|
| WTG | 12 |
| PV | 15 |
| Battery | 10 |

**Table 7.3** Data for the load demand

| Hour | Load demand |
|---|---|
| 1 | 31.83 |
| 2 | 31.4 |
| 3 | 31.17 |
| 4 | 31 |
| 5 | 31.17 |
| 6 | 32.1 |
| 7 | 32.97 |
| 8 | 34.1 |
| 9 | 37.53 |
| 10 | 38.33 |
| 11 | 40.03 |
| 12 | 41.17 |
| 13 | 39.67 |
| 14 | 41.7 |
| 15 | 42.1 |
| 16 | 41.67 |
| 17 | 40.7 |
| 18 | 40.07 |
| 19 | 38.63 |
| 20 | 36.4 |
| 21 | 34.1 |
| 22 | 32.8 |
| 23 | 32.5 |
| 24 | 32 |

The optimization model is formulated and solved using the CONPOT solver embedded in AIMMS. The AIMMS code for solving an example of a HRES with battery storage in a microgrid (HRES_B) problem is shown below:

```
Model Main_HRES_B {
  Set time {
    Index: t;
  }
  Set Gen {
    Index: i;
  }
  Index k {
    Range: time;
  }
  Parameter load {
    IndexDomain: (t);
  }
  Parameter wind {
    IndexDomain: (t);
  }
  Parameter PV {
    IndexDomain: (t);
  }
  Parameter Genmax {
    IndexDomain: (i);
  }
  Parameter Genmin {
    IndexDomain: (i);
  }
  Parameter CG;
  Parameter a {
    IndexDomain: (i);
  }
  Parameter b {
    IndexDomain: (i);
  }
  Parameter c {
    IndexDomain: (p);
  }
  Parameter bbb {
    IndexDomain: (p);
  }
  Parameter nc {
    Definition: 0.75;
  }
  Parameter nd {
    Definition: 1;
  }
  Parameter dod {
    Definition: 0.5;
  }
  Parameter Bmax {
    Definition: 50;
  }
  Parameter Bmin {
    Definition: 1;
    Comment: "(1-dod)*Bmax";
  }
  Variable Bc {
    IndexDomain: (t);
    Range: nonnegative;
  }
  Variable PSB {
    IndexDomain: (t);
    Range: nonnegative;
  }
```

```
Variable PB {
   IndexDomain: (t);
   Range: free;
}
Variable Pgen {
   IndexDomain: (i,t);
   Range: nonnegative;
}
Variable Tcost {
   Range: free;
   Definition: sum[(i,t),(a(i)+(b(i)*Pgen(i,t))+(c(i)*Pgen(i,t)^2))];
}
MathematicalProgram LeastCost {
   Objective: Tcost;
   Direction: minimize;
   Constraints: AllConstraints;
   Variables: AllVariables;
   Type: Automatic;
}
Constraint c1 {
   IndexDomain: (t);
   Definition: sum(i,Pgen(i,t))+ W(t)+PSB(t)=load(t);
   Comment: "sum(p,pch(p,t))+W(t)+PSB(t)=load(t)";
}
Constraint c3 {
   IndexDomain: (t);
   Definition: 0<=W(t)<=wind(t);
   Comment: "0<=W(t)<=wind(t)";
}
Constraint c4 {
   IndexDomain: (t);
   Definition: PSB(t)+PB(t)<=PV(t);
   Comment: "0<=S(t)<=PV(t)";
}
Constraint c5 {
   IndexDomain: (i,t);
   Definition: 0<=Pgen(i,t);
   Comment: "0<=Pgen(i,t)";
}
Constraint c6 {
   IndexDomain: (i,t);
   Definition: Pgen(i,t)<=Genmax(i);
   Comment: " Pgen(i,t)<=Genmax(i)";
}
Constraint bc1 {
   IndexDomain: (t);
   Definition: Bmin<=PB(t)<=Bmax;
   Comment: {
      "Pc(t)+S(t)<=PV(t); Bmin<=PB(t)<=Bmax"
   }
}
Constraint bc2 {
   IndexDomain: (t);
   Definition: Bc(t)=0.6-(nc*sum(k|ord(k) <=ord(t),PB(k)));
   Comment: "Bc(t)=Bc(t-1)+(nc*sum(k,Pc(t)))-(nd*sum(k,Pd(t)))";
}
Constraint bc3 {
   IndexDomain: (t);
   Definition: Bc(t)<=Bmax;
   Comment: "Bc(t)<=Bmax";
}
Constraint bc4 {
   IndexDomain: (t);
   Definition: Bmin<=PB(t);
   Comment: "Bmin<=PB(t)";
}
Procedure MainInitialization {
```

```
    Comment: "Add initialization statements here that do NOT require any library being initialized already.";
}
Procedure PostMainInitialization {
    Comment: {
        "Add initialization statements here that require that the libraries are already initialized properly,
        or add statements that require the Data Management module to be initialized."
    }
}
Procedure MainExecution {
    Body: {
        solve LeastCost;
    }
}
```

The simulated results are given below. Table 7.4 presents the optimal power produced by the generating units based on the daily operational demand. The power produced by the PV, wind, and battery system is presented in Figs. 7.1, 7.2 and 7.3. It can be observed that there is a substantial reduction in electricity production costs for the farming loads owing to increased use of PV, wind, and battery power.

**Table 7.4** Optimal power flow for the conventional and hybrid RES generating units

| Time (hr) | $P_i$(MW) | $B_{soc}$ | $P_B$(MW) | $P_v$(MW) | $P_w$(MW) |
|---|---|---|---|---|---|
| 1 | 18.87 | 4.65 | −5.39 | 5.39 | 7.56 |
| 2 | 18.5 | 8.41 | −5.02 | 5.02 | 7.5 |
| 3 | 17.89 | 11.45 | −4.04 | 4.04 | 8.25 |
| 4 | 18.47 | 14.18 | −3.64 | 3.64 | 8.48 |
| 5 | 19.04 | 17.04 | −3.81 | 3.81 | 8.48 |
| 6 | 18.86 | 19.89 | −3.80 | 3.80 | 9.42 |
| 7 | 19.34 | 23.10 | −4.27 | 4.27 | 9.82 |
| 8 | 19.47 | 20.76 | 5.12 | 4.87 | 10.35 |
| 9 | 21.77 | 18.67 | 4.79 | 7.77 | 10.88 |
| 10 | 19.54 | 14.79 | 8.17 | 8.44 | 11.01 |
| 11 | 20.64 | 11.23 | 6.76 | 10.21 | 10.94 |
| 12 | 20.27 | 8.69 | 8.39 | 11.61 | 10.68 |
| 13 | 17.63 | 5.38 | 5.41 | 10.37 | 10.42 |
| 14 | 21.17 | 3.95 | 3.92 | 12.67 | 10.15 |
| 15 | 19.75 | 3.94 | 2.01 | 13.55 | 9.67 |
| 16 | 19.13 | 5.43 | −1.98 | 13.81 | 8.98 |
| 17 | 18.51 | 7.89 | −3.28 | 13.45 | 8.37 |
| 18 | 19 | 12.33 | −5.92 | 13.58 | 7.61 |
| 19 | 18.34 | 22.12 | −10 | 10.05 | 6.7 |
| 20 | 17.62 | 30.98 | −8.80 | 8.80 | 5.72 |
| 21 | 15.08 | 36.99 | −7.01 | 7.01 | 7.21 |
| 22 | 17.03 | 41.62 | −6.17 | 6.17 | 7.75 |
| 23 | 18.44 | 45.92 | −5.74 | 5.74 | 7.88 |
| 24 | 18.56 | 50.00 | −5.43 | 5.43 | 7.69 |

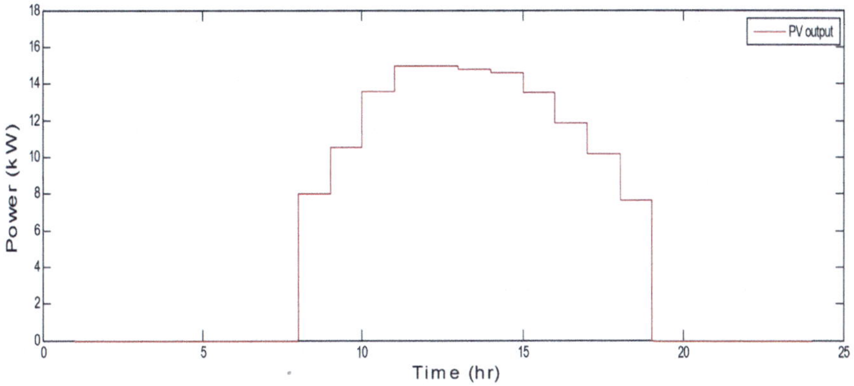


**Fig. 7.1** PV power generated

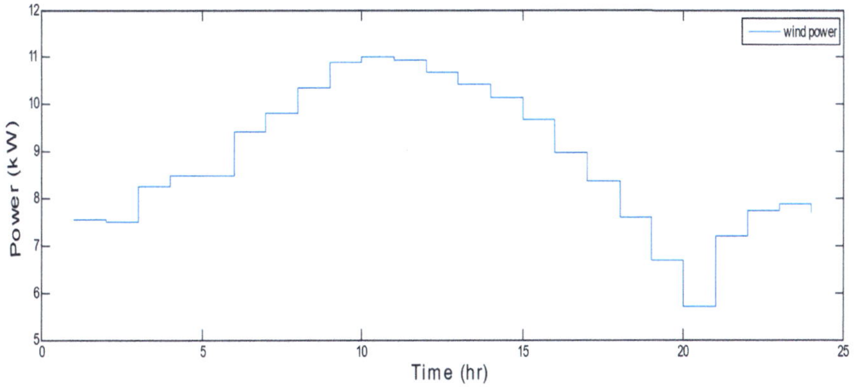


**Fig. 7.2** Wind power generated

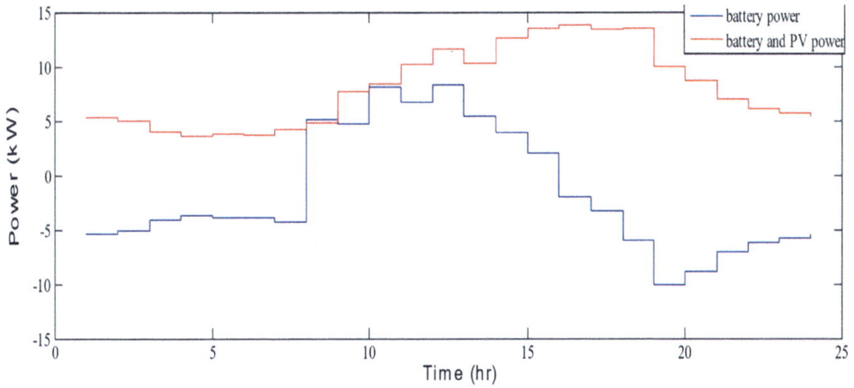


**Fig. 7.3** Power produced by PV and battery (charging and discharging) system

## 7.5 Incentive-Based Hybrid Renewable Energy Sources Powered Microgrid with Demand Response

This section discusses the energy management of microgrids, which is achieved through minimizing or maximizing some predetermined objectives such as cost, emission, and reliability, thereby determining the optimal commitment and economic schedule of the fossil fuel generators [22]. Recently, the customer side has been considered by providing demand relief on that side. This is referred to as DR. The inclusion of DR programs in microgrid energy management would ensure an efficient and more reliable microgrid, as this would adequately monitor the operating conditions of both the generation and demand sides of the system [23]. Furthermore, the incorporation of DR programs in a microgrid will reduce the operational cost of the microgrid and provide better grid flexibility. In this chapter, the DR program used is an incentive-based DR program, which ensures that the customers benefit by taking part in the DR program and are properly compensated based on their level of participation. We are considering a grid-connected microgrid operation system. There is an assumption that a trading system approach exists, giving the system operator the ability to sell or buy power from the grid owing to the variable nature of RES [17]. Therefore, the objective function is divided into two; the first is intended to minimize the operational cost of the fossil fuel generators and transaction cost for transferable power trading, while the second is meant to maximize customer DR benefits, as shown below:

$$\text{Min}\left[w_1\left(\sum_{t=1}^{T}\sum_{k=1}^{K}C_k\left(P_{k,t}\right)+\sum_{t=1}^{T}C_rP_{r,t}\right)+w_2\left(\sum_{t=1}^{T}\sum_{g=1}^{G}\left(z_{g,t}-\lambda_{g,t}x_{g,t}\right)\right)\right] \tag{7.24}$$

subject to the following constraints:

$$\sum_{k=1}P_{k,t}+P_{s,t}+P_{w,t}+P_{r,t}=d_t-\sum_{g=1}x_{g,t} \tag{7.25}$$

$$P_k^{\min}\leq P_{k,t}\leq P_k^{\max} \tag{7.26}$$

$$0\leq P_{s,t}\leq S_t \tag{7.27}$$

$$0\leq P_{w,t}\leq W_t \tag{7.28}$$

$$-\mathrm{DR}_k\leq P_{k,t+1}-P_{k,t}\leq \mathrm{UR}_k \tag{7.29}$$

$$-P_{r,t}^{\max} \leq P_{r,t} \leq P_{r,t}^{\max} \tag{7.30}$$

$$\sum_{t=1}\sum_{g=1} z_{g,t} \leq \Omega \tag{7.31}$$

$$\sum_{t=1} x_{g,t} \leq \mathrm{CM}_g \tag{7.32}$$

$$\sum_{t=1}\left[z_{g,t} - \left(q_{1,g}x_{g,t}^2 + q_{2,g}x_{g,t} - q_{2,g}x_{g,t}\theta_g\right)\right] \geq 0 \quad \text{for } g = 1, \ldots, G \tag{7.33}$$

$$\begin{aligned}&\sum_{t=1}\left[z_{g,t} - \left(q_{1,g}x_{g,t}^2 + q_{2,g}x_{g,t} - q_{2,g}x_{g,t}\theta_g\right)\right] \geq \\ &\sum_{t=1}\left[z_{g-1,t} - \left(q_{1,g-1}x_{g-1,t}^2 + q_{2,g-1}x_{g-1,t} - q_{2,g-1}x_{g-1,t}\theta_{g-1}\right)\right] \quad \text{for } g = 2, \ldots, G\end{aligned} \tag{7.34}$$

The desired variable results to be obtained by the optimization model are $z_{g,t}$, $x_{g,t}$, $P_{s,t}$, $P_{w,t}$, $P_{k,t}$ and $P_{r,t}$.

Equation (7.25) is the supply–demand constraint and it states that at any given time t, the total power produced by conventional, solar, and wind generating units and the transferrable power from the grid equals the load demand. Constraint (7.26) is the generating limit for a conventional generator. Constraints (7.27) and (7.28) are the generator limits for the solar and wind generators, respectively, which ensure that the solar and wind values do not exceed the forecast values. Constraint (7.29) is the ramp rates for the conventional generator. Constraint (7.30) is the power transferrable limit between the microgrid and the main grid. Constraint (7.31) is the utility budget limit and ensures that the total of the incentive paid is less than the budget. Equation (7.32) is the curtailment constraint, which states that on each customer's side, the total daily power curtailed must be less than the interruptible daily power limit. Constraint (7.33) states that the total of the customer incentive received must be greater or equal to the interruption costs daily. Constraint (7.34) states that the more power is curtailed, the greater the benefit to the customer.

The optimization model of the microgrid with DR is validated using three fossil fuel generators, one wind generator, one solar generator, and three rural customers. Table 7.5 presents parameters for fossil fuel generators. Table 7.6 gives the hourly load demand by the customer and interruptible power. The forecast wind and solar power obtained in [21] is presented in Table 7.7. The daily budget for the microgrid is \$500 (ZAR 8500) and the transferrable power is 4 kW.

**Table 7.5** Data for conventional generator

| Generator | $a_i$ | $b_i$ | $P_i^{min}$ | $P_i^{max}$ | $DR_i$ | $UR_i$ |
|---|---|---|---|---|---|---|
| 1 | 0.06 | 0.5 | 0 | 4 | 3 | 3 |
| 2 | 0.03 | 0.25 | 0 | 6 | 5 | 5 |
| 3 | 0.04 | 0.3 | 0 | 9 | 8 | 8 |

**Table 7.6** Hourly power demand and power interruptible values

| Hour | Load demand | Interruptible power |
|---|---|---|
| 1 | 31.83 | 1.57 |
| 2 | 31.4 | 1.4 |
| 3 | 31.17 | 2.2 |
| 4 | 31 | 3.76 |
| 5 | 31.17 | 4.5 |
| 6 | 32.1 | 4.7 |
| 7 | 32.97 | 5.04 |
| 8 | 34.1 | 5.35 |
| 9 | 37.53 | 6.7 |
| 10 | 38.33 | 6.16 |
| 11 | 40.03 | 6.38 |
| 12 | 41.17 | 6.82 |
| 13 | 39.67 | 7.3 |
| 14 | 41.7 | 7.8 |
| 15 | 42.1 | 8.5 |
| 16 | 41.67 | 7.1 |
| 17 | 40.7 | 6.8 |
| 18 | 40.07 | 6.3 |
| 19 | 38.63 | 5.8 |
| 20 | 36.4 | 4.2 |
| 21 | 34.1 | 3.8 |
| 22 | 32.8 | 3.01 |
| 23 | 32.5 | 2.53 |
| 24 | 32 | 1.42 |

**Table 7.7** Hourly power forecast from wind and solar generation

| Hour | Forecast wind power | Forecast solar power |
|---|---|---|
| 1 | 7.56 | 0 |
| 2 | 7.5 | 0 |
| 3 | 8.25 | 0 |
| 4 | 8.48 | 0 |
| 5 | 8.48 | 0 |
| 6 | 9.42 | 0 |
| 7 | 9.82 | 0 |
| 8 | 10.35 | 7.99 |
| 9 | 10.88 | 10.56 |
| 10 | 11.01 | 13.61 |
| 11 | 10.94 | 14.97 |
| 12 | 10.68 | 15 |
| 13 | 10.42 | 14.78 |
| 14 | 10.15 | 14.59 |
| 15 | 9.67 | 13.56 |
| 16 | 8.98 | 11.83 |
| 17 | 8.37 | 10.17 |
| 18 | 7.61 | 7.66 |
| 19 | 6.7 | 0 |
| 20 | 5.72 | 0 |
| 21 | 7.21 | 0 |
| 22 | 7.75 | 0 |
| 23 | 7.88 | 0 |
| 24 | 7.69 | 0 |

The optimization model is formulated and solved using the CONPOT solver embedded in AIMMS. The AIMMS code for solving an example of an incentive-based hybrid RES powered microgrid with DR (IBHRES) problem is shown below:

```
Model IBHRES {
  Set time {
    Index: t, tt;
  }
  Set customer {
    Index: g;
  }
  Set generator {
    Index: k;
  }
  Parameter pmin {
    IndexDomain: k;
  }
  Parameter pmax {
    IndexDomain: k;
  }
  Parameter a {
    IndexDomain: k;
  }
  Parameter b {
    IndexDomain: k;
  }
  Parameter ur {
    IndexDomain: k;
  }
  Parameter dr {
    IndexDomain: k;
  }
  Parameter tgencost {
    Definition: (sum[ (t,k), (b(k)*p(k,t))+(a(k)*p(k,t)*p(k,t))]);
  }
  Parameter totcustincentive {
    Definition: sum[ (t,g),z(g,t)];
  }
  Parameter totcustpower {
    Definition: sum[ (t,g),x(g,t)];
  }
  Parameter totpowgen {
    Definition: sum[ (k,t),p(k,t)];
  }
  Parameter d {
    IndexDomain: t;
  }
  Parameter thet {
    IndexDomain: g;
  }
  Parameter k1 {
    IndexDomain: g;
  }
  Parameter k2 {
    IndexDomain: g;
  }
  Parameter lam {
    IndexDomain: (g,t);
  }
  Parameter lamp {
    IndexDomain: t;
  }
  Parameter w {
    IndexDomain: (t);
  }
  Parameter s {
    IndexDomain: t;
  }
  Parameter w1 {
    Definition: 0.5;
  }
  Parameter w2 {
    Definition: 0.5;
  }
```

```
Parameter tcost {
   IndexDomain: g;
   Definition: sum[t,z(g,t)];
}
Parameter CM {
   IndexDomain: g;
}
Parameter tpow {
   IndexDomain: j;
   Definition: sum[t,x(g,t)];
}
Variable y {
   IndexDomain: (g,t);
   Range: nonnegative;
}
Variable pw {
   IndexDomain: t;
   Range: nonnegative;
}
Variable pr {
   IndexDomain: t;
   Range: (-4, 4);
}
Variable ps {
   IndexDomain: t;
   Range: nonnegative;
}
Variable u {
   IndexDomain: (k,t);
   Range: free;
}
Variable uu {
   IndexDomain: (g,t);
   Range: free;
}
Variable x {
   IndexDomain: (g,t);
   Range: nonnegative;

Variable p {
   IndexDomain: (k,t);
   Range: nonnegative;
}
Variable total {
   Range: free;
   Definition: (w1*(sum[ (t,k), (b(k)*p(k,t))+(a(k)*p(k,t)*p(k,t))]+sum[t,lamp(t)*( pr(t)) ]))+(w2*(sum[ (t,g), z(g,t)-
(lam(g,t)*x(g,t))]));
}
MathematicalProgram hmmm {
   Objective: total;
   Direction: minimize;
   Constraints: AllConstraints;
   Variables: AllVariables;
   Type: Automatic;
}
Constraint con1 {
   IndexDomain: (g,t);
   Definition: y(g,t)-((q1(g)*x(g,t)*x(g,t))+(q2(j)*x(g,t))-(q2(g)*x(g,t)*thet(g)))>=0;
}
Constraint con2 {
   IndexDomain: j;
   Definition: sum[t,(z(g,t)-((q1(g)*x(g,t)*x(g,t))+(q2(g)*x(g,t))-(q2(g)*x(g,t)*thet(g))))] >=sum[t,(z(g-1,t)-((q1(g-1)*x(g-
1,t)*x(g-1,t))+(q2(g-1)*x(g-1,t))-(q2(g-1)*x(g-1,t)*thet(g-1))))];
}
Constraint con4 {
   IndexDomain: t;
   Definition: sum[k, p(k,t)]+pw(t)+ps(t)+pr(t)=d(t)-sum[g, x(g,t)];
}
Constraint con5 {
   IndexDomain: (k,t);
   Definition: p(k,t)<=pmax(k);
}
```

```
Constraint con6 {
   IndexDomain: (k,t);
   Definition: p(k,t)>=pmin(k);
}
Constraint con7 {
   IndexDomain: (k,t) | 1<= ord(t) <=24;
   Definition: p(k,t)-p(k,t+1)<=dr(k);
}
Constraint con8 {
   IndexDomain: (k,t) | 1<= ord(t) <=24;
   Definition: p(k,t+1)-p(k,t)<=ur(k);
}
Constraint con9 {

   Definition: sum[t,sum[g, z(g,t)]]<=10000;
}
Constraint con10 {
   IndexDomain: g;
   Definition: sum[t,x(g,t)]<=CM(g);
}
Constraint con11 {
   IndexDomain: t;
   Definition: pw(t)=w(t);
}
Constraint con12 {
   IndexDomain: t;
   Definition: ps(t)=s(t);
}
Procedure MainInitialization;
Procedure MainExecution {
   Body: {
      solve hmmm;
      if(hmmm.ProgramStatus <> 'Optimal' ) then
      empty x,y,total;
      endif;
   }
}
```

The simulated results are presented below. Table 7.8 presents the power generated by fossil fuel generators and power transfer between the microgrid and the main grid. Figure 7.4 presents the optimal power output from the solar and wind generators. Figure 7.5 gives the power output produced by fossil fuel generators. Figure 7.6 shows the incentive received by the customer and the power curtailed.

**Table 7.8** Power generated from fossil fuel generators and transfer power

| Hour | $P_1$ | $P_2$ | $P_3$ | $P_r$ |
|---|---|---|---|---|
| 1 | 4 | 6 | 9 | 4 |
| 2 | 4 | 6 | 9 | 4 |
| 3 | 4 | 6 | 9 | 3.19 |
| 4 | 4 | 6 | 9 | −0.08 |
| 5 | 4 | 6 | 9 | −1.54 |
| 6 | 4 | 6 | 9 | −1.99 |
| 7 | 4 | 6 | 9 | −2.27 |
| 8 | 3.6 | 6 | 7.9 | −4 |

(continued)

**Table 7.8** (continued)

| Hour | $P_1$ | $P_2$ | $P_3$ | $P_r$ |
|---|---|---|---|---|
| 9 | 3.16 | 6 | 7.25 | −4 |
| 10 | 2.49 | 6 | 6.23 | −4 |
| 11 | 2.55 | 6 | 6.33 | −4 |
| 12 | 2.89 | 6 | 6.83 | −4 |
| 13 | 2.3 | 6 | 5.95 | −4 |
| 14 | 3.04 | 6 | 7.06 | −4 |
| 15 | 3.47 | 6 | 7.71 | −4 |
| 16 | 4 | 6 | 9 | −4 |
| 17 | 4 | 6 | 9 | −4 |
| 18 | 4 | 6 | 9 | −3.38 |
| 19 | 4 | 6 | 9 | 4 |
| 20 | 4 | 6 | 9 | 4 |
| 21 | 4 | 6 | 9 | 4 |
| 22 | 4 | 6 | 9 | 4 |
| 23 | 4 | 6 | 9 | 4 |
| 24 | 4 | 6 | 9 | 4 |

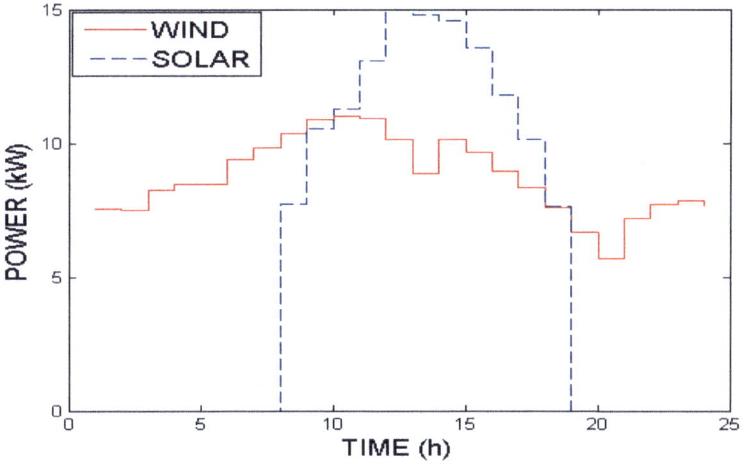


**Fig. 7.4** Wind and solar power

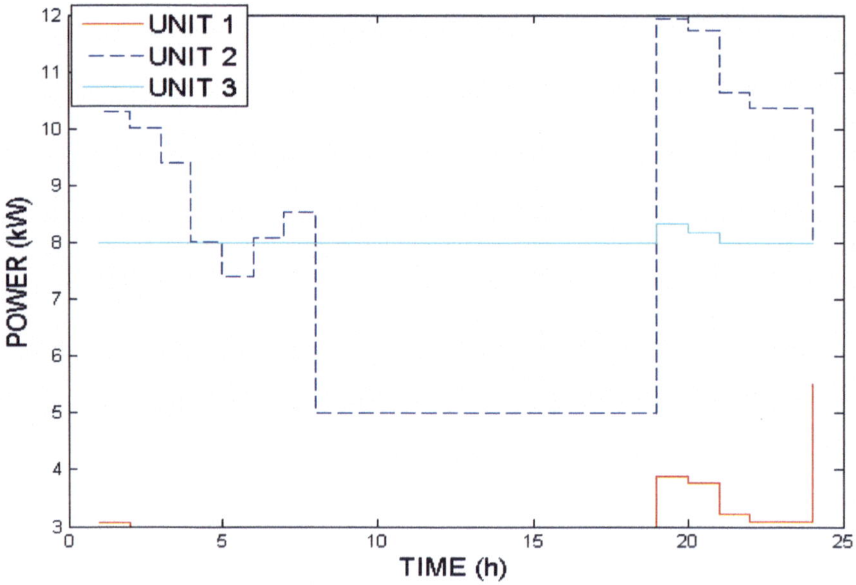


**Fig. 7.5** Power output from fossil fuel generators

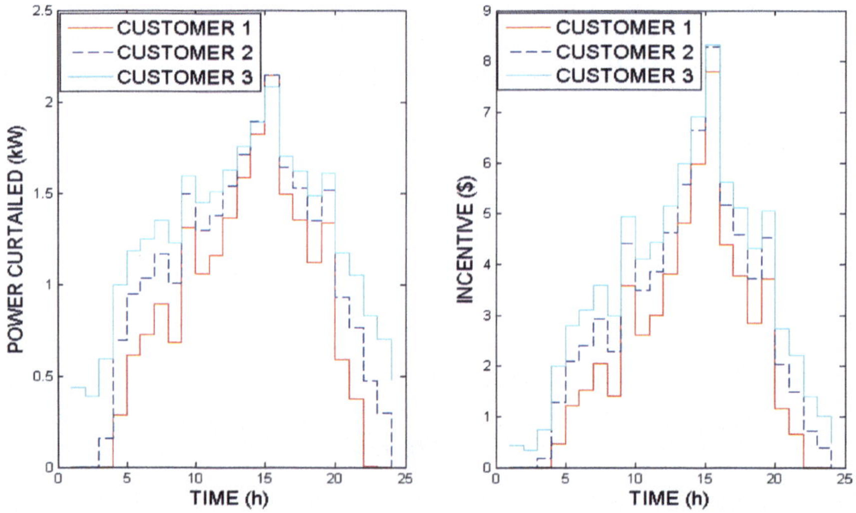


**Fig. 7.6** Incentive and power curtailed by customers

## 7.6 Summary

In this chapter, the operational mode of a grid-connected microgrid with incorporation of an HRES and a DR program was investigated. The main aim was to minimize the operational cost of the fossil fuel generators and the cost of transferrable power trading between the main grid and microgrid, as well as maximizing the incentive-based DR. The simulation results show better operational cost in the microgrid when considering the DR integration into the grid-connected microgrid against fuel and trading costs.

## 7.7 Nomenclature

The nomenclature used in this chapter is given below:

**Set**

| | |
|---|---|
| $B$ | battery |
| $g$ | customer |
| $k$ | generator |
| $s$ | solar |
| $t$ | time |
| $w$ | wind |

**Parameters**

| | |
|---|---|
| $C_k$ | fuel cost of generator $k$ |
| $C_r$ | transaction cost of transferable power |
| $d_t$ | total system load |
| $S_t$ | forecast solar power obtainable from solar generator at time $t$ |
| $W_t$ | forecast wind power obtainable from wind generator at time $t$ |
| $P_k^{\min}$ | minimum power capacity of generator $k$ |
| $P_k^{\max}$ | maximum power capacity of generator $k$ |
| $P_r^{\max}$ | maximum transferable power between the microgrid and grid |
| $\mathrm{DR}_k$ | maximum ramp-down rate for generator |
| $\mathrm{UR}_k$ | maximum ramp-up rate for generator $k$ |
| $\mathrm{CM}_g$ | interruptible power daily limit for customer $g$ |
| $\lambda_{g,t}$ | power interruptible value for different customer g at time $t$ |
| $\theta_g$ | type of customer |
| $\Omega$ | budget for the utility |
| $q_1$ and $q_2$ | coefficients of cost function for customer $g$ |
| $a_k$, $b_k$ and $c_k$ | coefficients for fuel cost for generator $k$ |

## Variables

$P_{k,t}$ power produced by generator $k$ at time $t$
$P_{B,t}$ charging and discharging power of the battery at time $t$
$P_{v,t}$ power flow from the PV and battery at time $t$
$P_{s,t}$ PV solar output power at time $t$
$P_{w,t}$ wind power output at time $t$
$P_{r,t}$ transferable power at time $t$
$x_{g,t}$ quantity of curtailed power by customer $g$ at time $t$
$z_{g,t}$ amount of incentive paid to customer $g$ at time $t$

# References

1. Dawoud SM, Lin X, Okba MI (2018) Hybrid renewable microgrid optimization techniques: a review. Renew Sustain Energy Rev 82:2039–2052
2. Gbadamosi SL, Nwulu NI, Sun Y (2018a) Multi-objective optimisation for composite generation and transmission expansion planning considering offshore wind power and feed-in tariffs. IET Renew Power Gener 12(14):1687–1697
3. Gbadamosi SL, Nwulu NI (2020) Optimal power dispatch and reliability analysis of hybrid CHP-PV-wind systems in farming applications. Sustainability 12
4. Gbadamosi SL, Nwulu NI, Sun Y (2018b) Harmonic and power loss minimization in power systems incorporating renewable energy sources and locational marginal pricing. J Renew Sustain Energy 10:1–24
5. Olatomiwa L, Blanchard R, Mekhilef S, Akinyele D (2018) Hybrid renewable energy supply for rural healthcare facilities: an approach to quality healthcare delivery. Sustain Energy Technol Assess 30:121–138
6. Babatunde OM, Denwigwe IH, Adedoja OS, Babatunde DE, Gbadamosi SL (2019) Harnessing renewable energy for sustainable agricultural applications. Int J Energy Econ Policy 9(5):308–315
7. Radhakrishnan BM, Srinivasan D (2016) A multi-agent based distributed energy management scheme for smart grid applications. Energy 103:192–204
8. Hajiaghasi S, Salemnia A, Hamzeh M (2019) Hybrid energy storage system for microgrids applications: a review. J Energy Storage 21:543–570
9. Gbadamosi SL, Nwulu NI (2020a) Optimal planning of renewable energy systems for power loss reduction in transmission expansion planning. J Eng Des Technol 18(5):1209–1222
10. Yang Y, Bremner S, Menictas C, Kay M (2018) Battery energy storage system size determination in renewable energy systems: a review. Renew Sustain Energy Rev 91:109–125
11. Gbadamosi SL, Nwulu NI (2019) Optimal microgrid sizing incorporating machine learning forecasting. In: International conference on industrial engineering and operations management, pp 1637–1642
12. Mousavi N, Kothapalli G, Habibi D, Das CK, Baniasadi A (2020) Modelling, design, and experimental validation of a grid-connected farmhouse comprising a photovoltaic and a pumped hydro storage system. Energy Convers Manag 210:112675
13. Mousavi N, Kothapalli G, Habibi D, Khiadani M, Das CK (2019) An improved mathematical model for a pumped hydro storage system considering electrical, mechanical, and hydraulic losses. Appl Energy 247:228–236

14. Mousavi SMG, Faraji F, Majazi A, Al-haddad K (2017) A comprehensive review of flywheel energy storage system technology. Renew Sustain Energy Rev 67:477–490
15. Arani AAK, Karami H, Gharehpetian GB, Hejazi MSA (2017) Review of flywheel energy storage systems structures and applications in power systems and microgrids. Renew Sustain Energy Rev 69:9–18
16. Yang B, Wang J, Sang Y, Yu L, Shu H, Li S (2019) Applications of supercapacitor energy storage systems in microgrid with distributed generators via passive fractional-order sliding-mode control. Energy 187:115905
17. Luta DN, Raji AK (2019) Optimal sizing of hybrid fuel cell-supercapacitor storage system for off-grid renewable applications. Energy 166:530–540
18. Gbadegesin AO, Sun Y, Nwulu NI (2019) Techno-economic analysis of storage degradation effect on levelised cost of hybrid energy storage systems. Sustain Energy Technol Assess 36:100536
19. Fortenbacher P, Demiray T (2019) Electrical power and energy systems linear/quadratic programming-based optimal power flow using linear power flow and absolute loss approximations. Electr Power Energy Syst 107:680–689
20. Nwulu NI, Xia X (2017) Optimal dispatch for a microgrid incorporating renewables and demand response. Renew Energy 101:16–28
21. Gbadamosi SL, Nwulu NI, Sun Y (2018) Optimal scheduling in offshore wind farms considering harmonic losses and locational marginal pricing. In: 2018 IEEE PES-IAS 5th power Africa conference, pp 658–663
22. Nwulu NI, Xia X (2015) Multi-objective dynamic economic emission dispatch of electric power generation integrated with game theory based demand response programs. Energy Convers Manag 89:963–974
23. Gbadamosi SL, Nwulu NI (2020b) A multi-period composite generation and transmission expansion planning model incorporating renewable energy sources and demand response. Sustain Energy Technol Assess 39:100726

# Chapter 8
# Optimal Power Flow

## 8.1 Introduction

This chapter provides information on an OPF optimization model to achieve a better operational state as well as system planning for effective service delivery to customers. OPF has become one of the most valuable optimization approaches in power system operation and planning; it aids generation scheduling to meet the required demand at lower operational cost while satisfying all necessary constraints [1, 2]. Basically, OPF models can be categorized into AC-OPF and DC-OPF models [3]. AC-OPF includes an optimization problem with a nonlinear and nonconvex nature, while DC-OPF is a linearized optimization model without involvement of voltage and reactive power in the model [4]. Here, we provide solutions for a DC-OPF model considering operational cost minimization and optimal generation scheduling in AIMMS.

## 8.2 DC Optimal Power Flow

DC-OPF is a linearized optimization model that is derived by simplifying the AC-OPF. It plays a key function in the operation and planning of modern power systems [4]. To perform as a simplified convex programming problem, the DC-OPF model ignores network losses and reactive power in the system. The following assumptions are made for DC-OPF [5, 6]:

- The resistance of the line is negligible because the reactance is much greater than the resistance (X >> R).
- The transmission line losses are negligible. Since $R = 0$, $I^2R$ losses are also ignored.
- The voltage magnitude at each bus is given as 1.0 per unit.
- The real power is greater than the reactive power, therefore the reactive power is negligible (P > Q). Hence, in DC-OPF, the reactive power flow equation does not exist.



N. Nwulu and S. L. Gbadamosi, *Optimal Operation and Control of Power Systems Using an Algebraic Modelling Language*, Green Energy and Technology,
https://doi.org/10.1007/978-3-030-00395-1_8

- The bus angle differences between the connected buses are less than 15°. Hence, $\sin(\theta_i - \theta_j)$ is given as zero and $\cos(\theta_i - \theta_j)$ is taken as $(\theta_i - \theta_j)$ in radians.

Therefore, the required variables of the DC-OPF optimization problem are the real output power of the generators and the phase angle difference between the connected buses. Generally, the DC-OPF analysis can be applied to contingency planning with a large-scale power system.

### 8.2.1 *Formulation of DC-OPF Model*

In this section, the objective function seeks to minimize the operational costs of the system while satisfying the physical and grid constraints.

#### 8.2.1.1 Objective Function

The DC-OPF model for the optimization problem can be formulated as:

$$\text{Min}\left[\sum_{g}\left(a_g + b_g P_g + c_g P_g^2\right)\right]. \tag{8.1}$$

Model Constraints

This is subject to the following model constraints:

$$P_g = P_d + (B_{ij}\theta_{ij}) \tag{8.2}$$

$$P_{ij} \leq P_{ij}^{\max} \tag{8.3}$$

$$P_g^{\min} \leq P_g \leq P_g^{\max} \tag{8.4}$$

$$\theta_{ij}^{\min} \leq \theta_{ij} \leq \theta_{ij}^{\max}. \tag{8.5}$$

The desired variable results to be obtained by the optimization model are $P_g$, $P_{ij}$, and $\theta_{ij}$.

- Equation (8.2) is the supply–demand balance constraint, which ensures that the power produced by the generators equals the demand and the power flowing on the transmission lines [7].
- Constraint (8.3) ensures that the power flowing on the transmission line does not exceed its maximum limit [8].
- Constraint (8.4) is the generator limit, which must be between the limit bounds [9].

- Constraint (8.5) is the voltage angle limit and must not exceed its maximum limit [10] (Table 8.1).

The AIMMS code for solving the example for a DC-OPF model is given below:

```
Model Main_14_Bus_System {
  Set Bus {
    Index: i;
  }
  Set Time {
    Index: t;
  }
  Set Buss {
    Index: j;
  }
  Parameter line {
    IndexDomain: (i,j);
    Range: binary;
  }
  Parameter LineFlowLim {
    IndexDomain: (i,j);
  }
  Parameter Ybus {
    IndexDomain: (i,j);
  }
  Parameter Pmax {
    IndexDomain: i;
  }
  Parameter Pmin {
    IndexDomain: i;
  }
  Parameter a {
    IndexDomain: i;
  }
  Parameter b {
    IndexDomain: i;
  }
  Parameter c {
    IndexDomain: i;
  }
  Parameter Pd {
```

```
    IndexDomain: i;
}
Parameter Ppv {
    IndexDomain: (i,t);
}
Variable LineFlow {
    IndexDomain: (i,j);
    Definition: Ybus(i,j)*Theta(i,j);
}
Variable NetFlow {
    IndexDomain: i;
    Definition: sum (j, LineFlow(i,j));
}
Parameter TotalGenOutput {
    Definition: sum (i, Pg(i));
}
Parameter TotalSystemDemand {
    Definition: sum (i, Pd(i));
}
Variable TotalSystemLosses {
    Range: free;
    Definition: sum (i, NetFlow(i));
}
Parameter check {
    IndexDomain: i;
    Definition: {

        Pg(i)=Pd(i)
    }
}
Variable Theta {
    IndexDomain: (i,j);
    Range: (-1.571, 1.571);
    Definition: line(i,j)*Theta(i,j);
}
Parameter Copy_check {
    IndexDomain: i;
    Definition: sum(j,Ybus(i,j)*Theta(i,j));
```

```
}
Variable Pg {
  IndexDomain: i;
  Range: nonnegative;
}
Variable Costt {
  Range: free;
  Definition: sum((i), a(i) + (b(i)*Pg(i))+(c(i)*Pg(i)*Pg(i)));
  Comment: "sum((i), a(i) + (b(i)*Pg(i))+(c(i)*Pg(i)*Pg(i)))";
}
Variable BusAng {
  IndexDomain: i;
  Range: (-0.78539, 0.78539);
}
MathematicalProgram LeastCost {
  Objective: Costt;
  Direction: minimize;
  Constraints: AllConstraints;
  Variables: AllVariables;
  Type: Automatic;
}
Constraint BusGenLowerLimit {
  IndexDomain: i;
  Definition: Pg(i)>=Pmin(i);
}
Constraint BusGenUpperLimit {
  IndexDomain: i;
  Definition: Pg(i)<=Pmax(i);
}
Constraint EnBal2 {
  Definition: sum(i, Pg(i))=sum(i, Pd(i)) + sum(i,sum(j,Ybus(i,j)*Theta(i,j)));
}
Constraint FlowLimit {
  IndexDomain: (i,j);
  Definition: Ybus(i,j)*Theta(i,j)<=LineFlowLim(i,j);
}
Constraint BusAng1 {
  IndexDomain: (i,j) |ord(i)>1;
```

```
    Definition: (BusAng(i)-0.78539)<=Theta(i,j);
}
Constraint BusAng2 {
    IndexDomain: (i,j) |ord(i)>1;
    Definition: Theta(i,j)<=(BusAng(i)+0.78539);
}
Constraint BusAng3 {
    IndexDomain: (i,j) |ord(i)=1;
    Definition: Theta(i,j)<=0.78539;
}
Constraint Copy_BusAng3 {
    IndexDomain: (i,j) |ord(i)=1;
    Definition: Theta(i,j)>=-0.78539;
}
Constraint Generation {
    Definition: sum(i,Pg(i))=sum(i,Pd(i)) + TotalSystemLosses;
}
Procedure MainInitialization {
    Comment: "Add initialization statements here that do NOT require any library being initialized already.";
}
Procedure PostMainInitialization {
    Comment: {
        "Add initialization statements here that require that the libraries are already initialized properly,
        or add statements that require the Data Management module to be initialized."
    }
}
Procedure MainExecution {
    Body: {
        solve LeastCost;
    }
}
```

**Table 8.1** Parameters for IEEE 14-bus test system

| Generator number | $P_i$ min (p.u.) | $P_i$ max (p.u.) | $a_i$ (\$/(MWh)$^2$) | $b_i$ (\$/MWh) | $c_i$ (\$/h) |
|---|---|---|---|---|---|
| G1 | 0.1 | 1.6 | 0.005 | 2.45 | 105.0 |
| G2 | 0.2 | 0.8 | 0.005 | 3.51 | 44.1 |
| G3 | 0.2 | 0.5 | 0.005 | 3.89 | 40.6 |
| Total system load demand | 2.593 | | | | |

**Table 8.2** Optimal power generated from the generators

| Generators | Buses | Optimal power (MW) |
|---|---|---|
| G1 | 1 | 160 |
| G2 | 2 | 67.70 |
| G3 | 3 | 29.70 |

**Table 8.3** Bus voltage angles

| Bus location | Bus angle |
|---|---|
| B2 | −0.78539 |
| B4 | −0.78539 |
| B5 | −0.78539 |
| B6 | −0.78539 |
| B7 | −0.78539 |
| B9 | −0.78539 |
| B10 | −0.78539 |
| B12 | −0.78539 |
| B13 | −0.78539 |

The results as obtained from the simulation are described in Tables 8.2, 8.3 and 8.4. Table 8.2 presents the optimal power dispatched from the three generators of the 14-bus system. The voltage angle of the buses is shown in Table 8.3 and the power flows on the line between two buses are presented in Table 8.4.

## 8.3 Summary

This chapter describes the OPF optimization formulation. The objective of this chapter is to minimize the operational cost of the power system while maintaining effective generation scheduling with efficient power flow on the transmission lines. The performance of the proposed model further affirms the possibility of applying the OPF problem to the expansion planning procedure.

## 8.4 Nomenclature

The nomenclature used in this chapter is given below:

### Set

$d$ demand
$i,j$ transmission line
$g$ generator.

**Table 8.4** Optimal power flow between two buses

| | B2 | B3 | B4 | B5 | B6 | B7 | B8 | B9 | B10 | B11 | B12 | B13 | B14 |
|---|---|---|---|---|---|---|---|---|---|---|---|---|---|
| B1 | −0.13 | | | −0.04 | | | | | | | | | |
| B2 | | −0.08 | −0.09 | −0.09 | | | | | | | | | |
| B4 | | | | −0.37 | | −0.07 | | −0.03 | | | | | |
| B5 | | | | | −0.06 | | | | | | | | |
| B6 | | | | | | | | | | −0.08 | −0.06 | −0.12 | |
| B7 | | | | | | | −0.09 | −0.14 | | | | | |
| B9 | | | | | | | | | −0.19 | | | | −0.06 |
| B10 | | | | | | | | | | −0.08 | | | |
| B12 | | | | | | | | | | | | −0.08 | |
| B13 | | | | | | | | | | | | | −0.05 |

### Parameter

$B_{ij}$ susceptance of transmission line *i*, *j*
$P_d$ total system load demand *d*
$P_{i,j}^{\max}$ maximum power flow on the transmission line *i*, *j*
$\theta_{i,j}^{\min}$ Minimum voltage angle on the transmission line *i*, *j*
$\theta_{i,j}^{\max}$ Maximum voltage angle on the transmission line *i*, *j*.

### Variables

$P_g$ power produced by generator *g*
$P_{i,j}$ power flow on the transmission line *i*, *j*
$\theta_{i,j}$ voltage angle on the transmission line *i*, *j*.

## References


1. Fortenbacher P, Demiray T (2019) Electrical power and energy systems linear/quadratic programming-based optimal power flow using linear power flow and absolute loss approximations. Electr Power Energy Syst 107:680–689
2. Galvani S, Marjani SR (2019) Optimal power fl ow considering predictability of power systems. Electr Power Syst Res 171:66–73
3. Gbadamosi SL, Nwulu NI (2020a) Optimal planning of renewable energy systems for power loss reduction in transmission expansion planning. J Eng Des Technol 18(5):1209–1222
4. Ding T, Li C, Li F, Chen T, Liu R (2017) A bi-objective DC-optimal power flow model using linear relaxation-based second order cone programming and its Pareto Frontier. Int J Electr Power Energy Syst 88:13–20
5. Zhang H (2013) Transmission expansion planning for large power systems. Arizona State University
6. Gbadamosi SL, Nwulu NI, Damisa U (2019) Impact of power losses on optimal expansion planning using quadratic programming. In: 6th IEEE international conference on engineering technologies and applied sciences (ICETAS), pp 1–4
7. Gbadamosi SL, Nwulu NI, Sun Y (2018a) Multi-objective optimisation for composite generation and transmission expansion planning considering offshore wind power and feed-in tariffs. IET Renew Power Gener 12(14):1687–1697
8. Gbadamosi SL, Nwulu NI, Sun Y (2018b) Harmonic and power loss minimization in power systems incorporating renewable energy sources and locational marginal pricing. J Renew Sustain Energy 10:1–24
9. Gbadamosi SL, Nwulu NI (2020b) A multi-period composite generation and transmission expansion planning model incorporating renewable energy sources and demand response. Sustain Energy Technol Assess 39:100726
10. Gbadamosi SL, Nwulu NI (2020) Reliability assessment of composite generation and transmission expansion planning incorporating renewable energy sources. J Renew Sustain Energy 026301(12)

# Chapter 9
# Transmission Network Expansion Planning

## 9.1 Introduction

This chapter provides analysis of the TNEP problem with the aim of assisting system planners to recognize the necessary positions for adequate reinforcement and meeting the requirement for expansion planning. TNEP in a modern power system is an important research area because transmission lines aid the supply of power generated by the utilities at one end to the consumers at the other end. To ensure continuous supply to meet the growing demand, adequate reinforcement of lines must be devised with their locations and capacity, with a view to improving the system's reliability and economics on the power system.

TNEP is generally regarded as a long-term planning horizon because it is capital-intensive and has a long-lasting effect on the power system. There are two expansion planning strategies, namely, a static and a dynamic model. Static TNEP (STNEP) makes expansion plans at the start of the planning horizon taking annual investment costs into account, while dynamic TNEP (DTNEP) involves establishing expansion planning decisions at several stages of the planning period using annual representation of investment decisions [1]. From the perspective of modeling, there are two modeling approaches to the TNEP problem, namely, the DC model and the AC model. The DC model has the attribute of easy computation of power flow and convergence of the solutions is straightforward. Though AC power flow has adequate accuracy, it is more complicated and complex when compared to the DC model. Here, we provide solutions for static and dynamic TNEP problems for both DC and AC models in AIMMS.



N. Nwulu and S. L. Gbadamosi, *Optimal Operation and Control of Power Systems Using an Algebraic Modelling Language*, Green Energy and Technology,
https://doi.org/10.1007/978-3-030-00395-1_9

## 9.2 Static Transmission Network Expansion Planning

In this section, a static planning approach is considered for TNEP models and the expansion planning is determined for a short-time planning horizon [2, 3]. The mathematical formulation is presented below; it seeks to minimize the investment cost for new transmission lines and the operational cost of the generating units in the system.

### *9.2.1 DC Model*

The DC model for the optimization problem can be formulated as:

$$\mathrm{Min}\left[\sum_{l=1}^{n_l} C_l^i z_l + \beta \sum_{g=1}^{n_g} C_g^o P_g\right] \tag{9.1}$$

$$C_g^o P_g = \alpha_g + \beta_g P_g + \delta_g P_g^2 \tag{9.2}$$

subject to the following constraints:

$$\sum_g P_g - \sum_{l \in sl} P_l^+ + \sum_{r \in rl} P_l^- = \sum_k P_k \quad \forall b \tag{9.3}$$

$$\sum_l C_l^i z_l \le UB_l \tag{9.4}$$

$$z_l \in [0, 1] \quad \forall l \in pl \tag{9.5}$$

$$P_l = B_l(\phi_{l \in sl} - \phi_{l \in rl}) \quad \forall el \tag{9.6}$$

$$-(1 - z_l)M \le P_l - B_l(\phi_{l \in sl} - \phi_{l \in rl}) \le M(1 - z_l) \quad \forall pl \tag{9.7}$$

$$-P_l^{\max} \le P_l \le P_l^{\max} \quad \forall el \tag{9.8}$$

$$-z_l P_l^{\max} \le P_l \le z_l P_l^{\max} \quad \forall pl \tag{9.9}$$

$$P_g^{\min} \le P_g \le P_g^{\max} \quad \forall g \tag{9.10}$$

$$-\pi \le \phi_b \le \pi \quad \forall b. \tag{9.11}$$

The desired variable results to be obtained by the optimization model are $z_l$, $\phi_b$, $P_g$ and $P_l$.

The main objective of TNEP is to minimize the costs accrued in constructing new transmission lines. To ensure the system's reliability and avoid violation of the economic motives, the objective functions are bounded by the operational constraints.

The objective functions in Eq. (9.1) use two terms, namely:

(a) $\sum_{l=1}^{n_l} C_l^i z_l$ represents the annualized cost of constructing new transmission lines.
(b) $\sum_{g=1}^{n_g} C_g^o P_g$ represents the cost of operating the generating units.

Since the investment cost is annualized, the duration factor is given as 8760 h [4]. Therefore, to ensure comparable objectives, the duration factor is multiplied with the second objective.

For simplicity purposes, a quadratic cost function is applied for the fuel cost functions of the generating units [5]. In Eq. (9.2), the fuel cost in $/h is given as a function of its output power [2]. Constraint (9.3) ensures that at any bus b, the capacity sum of the power from generating units and the power flowing in and out of the transmission lines equals the load demand. Equation (9.4) is the constraint for the investment budget for the new transmission lines to be built and it states that the costs of constructing new additional lines must not exceed the available funds. Constraint (9.5) is the binary decision variable for the candidate transmission lines to be built and it states that the variable must be equal to 1 if a newly added line is built and 0 if otherwise [6]. Constraints (9.6) and (9.7) define the power flow on the existing and prospective transmission lines, respectively, and this is a function of the line susceptance $(B_l)$ and bus voltage angle at the sending and receiving end of the buses. Constraints (9.8) and (9.9) ensure that the line capacity for existing and prospective lines does not exceed its maximum limit. Constraint (9.10) ensures that the power from the generating units is within the minimum and maximum generating limits. Constraint (9.11) ensures that the phase angle at each bus is kept within the safe limit.

The IEEE Garver six-bus system data for the generators and the demand are given in Table 9.1 and the diagram is depicted in Fig. 9.1. This example contains five buses with three generating units, five load points, and six existing transmission lines connecting these buses. The sixth bus is isolated from the rest of the buses, but new lines can be built to connect to the rest of the buses if needed (Table 9.2).

**Table 9.1** Data for IEEE Garver six-bus system example

| Bus | Generation capacity (MW) | Plant type | $\alpha_g$ | $\beta_g$ | $\delta_g$ | $a_g$ | $b_g$ | $c_g$ | Demand (MW) |
|---|---|---|---|---|---|---|---|---|---|
| 1 | 150 | Fossil fuel | 200 | 11 | 0.009 | 40.2669 | −0.5455 | 0.00683 | 160 |
| 2 | – | – | – | – | – | – | – | – | 480 |
| 3 | 360 | Fossil fuel | 220 | 8.5 | 0.009 | 40.2669 | −0.5455 | 0.00683 | 80 |
| 4 | – | – | – | – | – | – | – | – | 320 |
| 5 | – | – | – | – | – | – | – | – | 480 |
| 6 | 600 | Hydro | – | – | – | – | – | – | – |

**Fig. 9.1** IEEE Garver six-bus system network

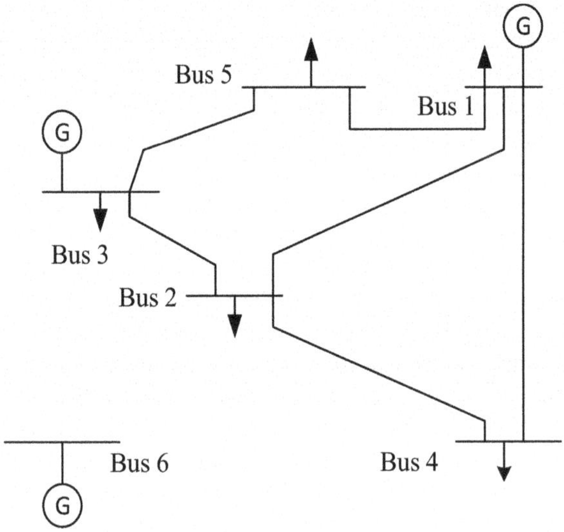


**Table 9.2** Parameters and capacities of the existing and prospective transmission lines

| Between buses | Line capacity (MW) | Susceptance, $B_L$ | Investment cost ($M) |
|---|---|---|---|
| 1–2 | 100 | 250 | |
| 1–4 | 80 | 133 | |
| 1–5 | 100 | 500 | |
| 2–3 | 100 | 250 | |
| 2–4 | 100 | 500 | |
| 3–5 | 100 | 250 | |
| 2–6 | 100 | 250 | 20 |
| 2–6 | 100 | 333 | 31 |
| 3–5 | 100 | 333 | 30 |
| 4–6 | 100 | 500 | 30 |
| 4–6 | 100 | 500 | 30 |

The AIMMS code for solving the example for a DC model of STNEP is given below:

```
Model Main_DC-STNEP
Sets
  Set Buses
    Index: n;
  Set Gen
    Index: g;
  Set Lines
    Index: L;
  Set existinglines
    SubsetOf: Lines;
  Set prospectivelines
    SubsetOf: Lines;
  Set Demand
    Index: k;
Parameters
  Parameter CombineGen
    IndexDomain: (g,n);
  Parameter CombineDemand
    IndexDomain: (k,n);
  Parameter Combinesl
    IndexDomain: (L,n);
  Parameter Combinerl
    IndexDomain: (L,n);
  Parameter LineCap
    IndexDomain: (L);
  Parameter LineSusceptance
    IndexDomain: (L);
  Parameter GenCap
    IndexDomain: (g);
  Parameter Costa
    IndexDomain: (g);
  Parameter Costb
    IndexDomain: (g);
  Parameter Costc
    IndexDomain: (g);
  Parameter Load
    IndexDomain: (k);
  Parameter LinvstCost
    IndexDomain: (L);
  Parameter Budget;
  Parameter M;
  Parameter Duration;
Variables
  Variable ZL
    IndexDomain: (L);
    Range: binary;
  Variable GenP
    IndexDomain: (g);
    Range: nonnegative;
  Variable Flow
    IndexDomain: (L);
    Range: nonnegative;
  Variable VoltAngle
      IndexDomain: (n);
      Range: free;
  Objective Function
    Variable TotalCost
      Range: free;
      Definition:                    sum(L|                    prospectivelines,
LinvstCost(L)*ZL(L))+(Duration*(sum(g,(Costa(g)+Costb(g)*GenP(g)+Costc(g)*GenP(g)^2))));
 MathematicalProgram NetCost
      Objective: TotalCost;
      Direction: minimize;
      Constraints: AllConstraints;
        Variables: AllVariables;
        Type: Automatic;
```

```
Constraints
  Constraint Nodalbalance
    IndexDomain: (n);
    Definition:                                                    sum(g|CombineGen(g,n),GenP(g))-
sum(L|Combinesl(L,n),Flow(L))+sum(L|Combinerl(L,n),Flow(L))=sum(k|CombineDemand(k,n),Load(d));
  Constraint Budgetlimit
    Definition: sum(L|prospectivelines,LinvstCost(L)*ZL(L))<=Budget;
  Constraint GenLimit
    IndexDomain: (g);
    Definition:
      0<=GenP(g)<=GenCap(g);
  Constraint AngleLimit
    IndexDomain: (n);
    Definition: -3.14<=VoltAngle(n)<=3.14;
  Constraint LineLimit1
    IndexDomain: (L)|ord(L)<=6;
    Definition: -LineCap(L)<=Flow(L);
  Constraint LineLimit2
    IndexDomain: (L)|ord(L)<=6;
    Definition: Flow(L)<=LineCap(L);
  Constraint LineLimit3
    IndexDomain: (L)|ord(L)>6;
    Definition:
      -LB(L)*LineCap(L)<=Flow(L);
  Constraint LineLimit4
    IndexDomain: (L)|ord(L)>6;
    Definition:
      LB(L)*LineCap(L)>=Flow(L);
  Constraint FlowLimit1
    IndexDomain: (L)|ord(L)<=6;
    Definition:
      Flow(L)=LineSusceptance(L)*(sum(n|Combinesl(L,n),VoltAngle(n))-sum(n|Combinerl(L,n),VoltAngle(n)));
  Constraint FlowLimit2
    IndexDomain: (L)|ord(L)>6;
    Definition:
      -(1-ZL(L))*M<=Flow(L)-LineSusceptance(L)*(sum(n|Combinesl(L,n),VoltAngle(n))-
sum(n|Combinerl(L,n),VoltAngle(n)));
  Constraint FlowLimit3
    IndexDomain: (L)|ord(L)>6;
    Definition:                        Flow(L)-LineSusceptance(L)*(sum(n|Combinesl(L,n),VoltAngle(n))-
sum(n|Combinerl(L,n),VoltAngle(n)))<=(1-ZL(L))*M;

  Procedure MainExecution {
    Body: {
      solve NetCost;
```

The optimization results as obtained from simulation are given in Tables 9.3, 9.4 and 9.5. It is clearly seen from Table 9.4 that five new transmission lines are added to the system network, four of which are connected to the isolated bus 6. Only three transmission lines are operating at full capacity (Fig. 9.2).

**Table 9.3** Optimal generated power from the generators for DC STNEP model

| Generators | Buses | Optimal power (MW) |
|---|---|---|
| G1 | 1 | 146.2936 |
| G2 | 3 | 337.1629 |
| G3 | 6 | 276.5435 |

**Table 9.4** Optimal power flow on the existing and prospective transmission lines for DC STNEP model

| Between buses | Optimal flow |
|---|---|
| 1–2 | 17.16 |
| 1–4 | 9.13 |
| 1–5 | 40 |
| 2–3 | 45 |
| 2–4 | 97.16 |
| 3–5 | 50.24 |
| 2–6 | 75.43 |
| 2–6 | 75.43 |
| 3–5 | 75.43 |
| 4–6 | 100 |
| 4–6 | 100 |

**Table 9.5** Voltage phase angle at the buses for DC STNEP model

| Buses | Phase angle (Rad) |
|---|---|
| 1 | 2.82 |
| 2 | 2.751348 |
| 3 | 3.14 |
| 4 | 2.751348 |
| 5 | 2.74 |
| 6 | 2.902218 |

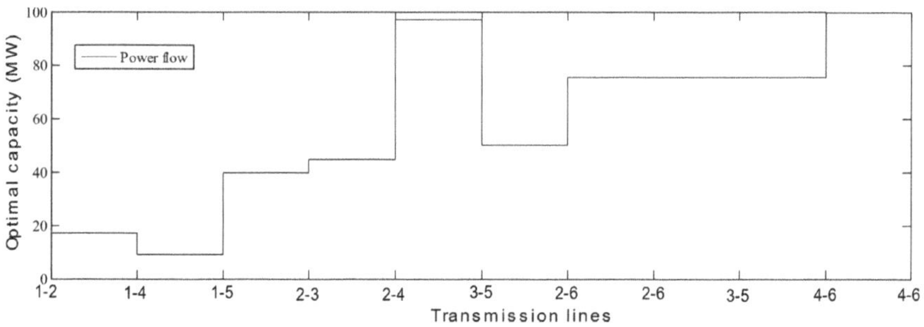


**Fig. 9.2** Optimal power flowing on the existing and prospective transmission lines

### 9.2.2 Nomenclature

**Set**

$g$ generator
$k$ demand
$l$ transmission line

**Parameters**

| | |
|---|---|
| $C_g^o$ | operational cost of the existing generator $g$ |
| $E_g$ | emissions cost of the existing generator $g$ |
| $\beta$ | duration factor |
| $P_k$ | system load demand $k$ |
| $P_l^{\max}$ | maximum power capacity for the transmission line $l$ |
| $P_g^{\min}$ and $P_g^{\max}$ | minimum and maximum power capacity of the generator $g$ |
| $UB_l$ | utility budget for prospective transmission line |
| $B_l$ | susceptance of transmission line |

**Variables**

| | |
|---|---|
| $P_g$ | power produced by the existing generator $g$ |
| $P_l$ | power flowing on transmission line $l$ |
| $P_l^+$ and $P_l^-$ | power flowing into and out of the transmission line $l$ |
| $z_l$ | binary variable for prospective transmission $l$ |
| $\phi_b$ | voltage angle at bus $b$ |

### 9.2.3 AC Model

The AC model for the optimization problem can be formulated as:

$$\text{Min}\left[\sum_{l=1}^{n_l} C_l^i z_l + \beta \sum_{g=1}^{n_g} C_g^o P_g\right] \tag{9.12}$$

$$C_g^o P_g = \alpha_g + \beta_g P_g + \delta_g P_g^2 \tag{9.13}$$

subject to the following constraints:

$$\sum_{g} P_g - \sum_{l \in sl} P_l^+ + \sum_{l \in rl} P_l^- = \sum_{k} P_k \quad \forall b \tag{9.14}$$

$$\sum_{g} Q_g - \sum_{l \in sl} Q_l^+ + \sum_{l \in rl} Q_l^- = \sum_{k} Q_k \quad \forall b \tag{9.15}$$

$$\sum_{l} C_l^i z_l \leq UB_l \tag{9.16}$$

$$z_l \in [0, 1] \quad \forall l \in pl \tag{9.17}$$

$$P_l^+ = V_a g_l - V_a V_b (g_l \cos \theta_{ab} + \mathrm{b}_l \sin \theta_{ab}) \quad \forall l \tag{9.18}$$

$$Q_l^+ = -V_a(b_l + b_{lo}) + V_a V_b (g_l \cos \theta_{ab} - b_l \sin \theta_{ab}) \quad \forall l \tag{9.19}$$

$$P_l^- = V_b g_l - V_a V_b (g_l \cos \theta_{ab} - b_l \sin \theta_{ab}) \quad \forall l \tag{9.20}$$

$$Q_l^- = -V_b(b_l + b_{lo}) + V_a V_b (b_l \cos \theta_{ab} + g_l \sin \theta_{ab}) \quad \forall l \tag{9.21}$$

$$P_{\text{loss}} = P_l^+ + P_l^- = g_l(V_a^2 + V_b^2 - 2V_a V_b \cos \theta_{ab}) \quad \forall l \tag{9.22}$$

$$Q_{\text{loss}} = Q_l^+ + Q_l^- = -b_{lo}(V_a^2 + V_b^2) - b_l(V_a^2 + V_b^2 - 2V_a V_b \cos \theta_{ab}) \quad \forall l \tag{9.23}$$

$$P_g^{\min} \leq P_g \leq P_g^{\max} \quad \forall g \tag{9.24}$$

$$Q_g^{\min} \leq Q_g \leq Q_g^{\max} \quad \forall g \tag{9.25}$$

$$0 \leq (P_l^+)^2 + (Q_l^+)^2 \leq S_l^2 \quad \forall l \tag{9.26}$$

$$0 \leq (P_l^-)^2 + (Q_l^-)^2 \leq S_l^2 \quad \forall l \tag{9.27}$$

$$V_{\min} \leq V_b \leq V_{\max} \quad \forall b \tag{9.28}$$

$$-\pi \leq \phi_b \leq \pi \quad \forall b. \tag{9.29}$$

**Table 9.6** Real and reactive parameters for IEEE Garver six-bus system example

| Bus | Parameters of generators | | | | Load parameter | |
|---|---|---|---|---|---|---|
| | $P_g^{min}$ (MW) | $P_g^{max}$ (MW) | $Q_g^{min}$ (MVAr) | $Q_g^{max}$ (MVAr) | $P_k$ (MW) | $Q_k$ (MVAr) |
| 1 | 0 | 160 | −10 | 65 | 80 | 16 |
| 2 | – | – | – | – | 240 | 48 |
| 3 | 0 | 360 | −10 | 150 | 40 | 8 |
| 4 | – | – | – | – | 160 | 32 |
| 5 | – | – | – | – | 240 | 48 |
| 6 | 0 | 600 | −10 | 200 | – | – |

The desired variable results to be obtained by the optimization model are $z_l$, $\phi_b$, $V_b$, $P_g$, $P_l^+$, $Q_l^+$, $P_l^-$, $Q_l^-$, $P_{\text{loss}}$ and $Q_{\text{loss}}$.

In this section, the objective function is the same as obtained in Sect. 9.2.1. The optimization problem seeks to minimize the investment cost of new transmission lines and the operational cost of the generating units simultaneously. The AC model for STNEP was validated on the IEEE Garver six-bus test system, as shown in Fig. 9.1, and the data are given in Table 9.6 as obtained in [7].

Equation (9.2) is the same as Eq. (9.13), representing the quadratic cost function for operational cost for the generators. Equations (9.14) and (9.15) are the real and reactive power balance constraints for the optimization problem developed. Constraints (9.16) and (9.17) are the same as given in (9.4) and (9.5) as the budgetary and binary decision variable constraints, respectively. Equations (9.18)–(9.21) are the constraints for the real and reactive power flowing out of and into the transmission lines, respectively. Constraints (9.22) and (9.23) are the real and reactive power loss on the transmission lines and these are derived from Eqs. (9.18) to (9.21). These constraints ensure that the sum of the power flowing out of and into the transmission line is equal to the real and reactive power loss. Constraints (9.24) and (9.25) stand for the real and reactive generation limit. Constraints (9.26) and (9.27) are the real and reactive transmission line limits. Lastly, the bus voltage magnitude and phase angle constraints are given in Eqs. (9.28) and (9.29), respectively [8].

The AIMMS code for solving the example for AC model of STNEP is given below:

```
Model Main_AC-STNEP
Set
  Set Nodes
    Index: n;
  Set Lines
    Index: L;
  Set Gen
    Index: g;
  Set Demand
    Index: k;
  Set existinglines
    SubsetOf: Lines;
  Set prospectivelines
    SubsetOf: Lines;
  Set RefNode
    SubsetOf: Nodes;
  Parameter CombineGen
    IndexDomain: (g,n);
  Parameter CombineDemand
    IndexDomain: (k,n);
  Parameter Combine_sl
    IndexDomain: (L,n);
  Parameter Combine_rl
    IndexDomain: (L,n);
  Parameter Pk
    IndexDomain: (k);
  Parameter Qk
    IndexDomain: (k);
  Parameter Pgmax
    IndexDomain: (g);
  Parameter Qgmax
    IndexDomain: (g);
  Parameter Qgmin
    IndexDomain: (g);
  Parameter LineCap
    IndexDomain: (L);
  Parameter Linvestcost
    IndexDomain: (L);
  Parameter Duration;
  Parameter LineSusc
    IndexDomain: (L);
  Parameter LineSusc2
    IndexDomain: (L);
  Parameter Conductance
    IndexDomain: (L);
  Parameter Vmin
    IndexDomain: (n);
  Parameter Vmax
    IndexDomain: (n);
Objective function
    Definition:
      sum(L,Conduct(L)*(sum(n|Combine_sl(L,n),sqr(Voltage(n)))+sum(n|Combine _rl(L,n),sqr(Voltage(n)))-
2*(sum(n|Combine _sl(L,n),Voltage(n))*sum(n|Combine _rl(L,n),Voltage(n))*cos(sum(n|Combine _sl(L,n),Angle(n))-
sum(n|Combine _rl(L,n),Angle(n))))));
  Parameter RL
    IndexDomain: (L)
  Parameter XL
    IndexDomain: (L);
  Parameter Costa
    IndexDomain: (g);
  Parameter Costb
    IndexDomain: g;
  Parameter Costc
    IndexDomain: g;
Variables
  Variable RFL_SN
```

```
    IndexDomain: (L);
    Range: nonnegative;
  Variable QFL_SN
    IndexDomain: (L);
    Range: free;
  Variable RFL_RN
    IndexDomain: (L);
    Range: free;
  Variable QFL_RN
    IndexDomain: (L);
    Range: free;
  Variable ZL
    IndexDomain: (L);
    Range: binary;
  Variable Voltage
    IndexDomain: (n);
    Range: nonnegative;
  Variable Pg
    IndexDomain: (g);
    Range: nonnegative;
  Variable Qg
    IndexDomain: (g);
    Range: free;
  Variable PLoss
    IndexDomain: (L);
    Range: free;
  Variable QLoss
    IndexDomain: (L);
    Range: free;
  Variable Voltangle
    IndexDomain: (n);
    Range: free;
  Variable Totalcost
    Range: free;
    Definition:
sum(L|prospectivelines,Linvestcost(L)*ZL(L))+(duration*(sum(g,(Costa(g)+Costb(g)*Pg(g)+Costc(g)*Pg(g)^2))));
  }
  MathematicalProgram NetCost
    Objective: Totalcost;
    Direction: minimize;
    Constraints: AllConstraints;
    Variables: AllVariables;
    Type: Automatic;
  ElementParameter myGMP
    Range: AllGeneratedMathematicalPrograms;
  Constraint nodalbalance1
    IndexDomain: n;
    Definition:
      sum(g|          CombineGen(g,n),Pg(g))-sum(L|          Combine_sl(L,n),RFL_SN(L))-sum(L|
Combine_rl(L,n),RFL_RN(L))=sum(k| CombineDemand(k,n),Pk(k));
  Constraint nodalbalance2
    IndexDomain: n;
    Definition:     sum(g|GenMap(g,n),Qg(g))-sum(L|Mapsl(L,n),QFL_SN(L))-sum(L|Maprl(L,n),QFL_RN(L))=sum(k|
CombineDemand(dk,n),Qk(k));
  Constraint Voltlimit1
    IndexDomain: n;
    Definition:
      Vmin(n)<=Voltage(n);
  Constraint Voltlimit2
    IndexDomain: n;
    Definition:
      Voltage(n)<=Vmax(n);
  Constraint Anglelimit1
    IndexDomain: n;
    Definition:
      -3.14<=Angle(n);
```

```
Constraint Anglelimit2
  IndexDomain: n;
  Definition: Angle(n)<=3.14;
Constraint Genlimit1
  IndexDomain: g;
  Definition:
    0<=Pg(g);
Constraint Genlimit2
  IndexDomain: g;
  Definition:
    Pg(g)<=Pgmax(g);
Constraint Genlimit3
  IndexDomain: g;
  Definition:
    Qgmin(g)<=Qg(g);
Constraint Genlimit4
  IndexDomain: g;
  Definition: Qg(g)<=Qgmax(g);
Constraint Flowlimit1
  IndexDomain: (L);
  Definition: RFL_SN(L)=(sqr(sum(n| Combine_sl(L,n),Voltage(n)))*Conductance(L))-
(sum(n|Mapsl(L,n),Voltage(n))*sum(n| Combine_rl(L,n),Voltage(n)))*(Conductance(L)*(cos(sum(n|
Combine_sl(L,n),Angle(n))-sum(n| Combine_rl(L,n),Angle(n))))+(-LineSusceptance(L)*sin(sum(n|
Combine_sl(L,n),Angle(n))-sum(n| Combine_rl(L,n),Angle(n)))));
Constraint Flowlimit2
  IndexDomain: L;
  Definition: QFL_SN(L)=-(sqr(sum(n|Mapsl(L,n),Voltage(n))))*(LineSusc2(L)+(-
LineSusceptance(L)))+((sum(n|Mapsl(L,n),Voltage(n))*sum(n|Maprl(L,n),Voltage(n)))*(((-
LineSusceptance(L))*(cos(sum(n|Mapsl(L,n),Angle(n))-sum(n|Maprl(L,n),Angle(n)))))-
(Conductance(L)*(sin(sum(n|Mapsl(L,n),Angle(n))-sum(n|Maprl(L,n),Angle(n)))))));
Constraint Flowlimit3
  IndexDomain: L;
  Definition: RFL_RN(L)=(sqr(sum(n| Combine_prl(L,n),Voltage(n)))*Conductance(L)-((sum(n|
Combine_sl(L,n),Voltage(n))*sum(n| Combine_rl(L,n),Voltage(n)))*(Conductance(L)*(cos(sum(n|
Combine_sl(L,n),Angle(n))-sum(n| Combine_rl(L,n),Angle(n)))))-(-LineSusceptance(L)*sin(sum(n|
Combine_sl(L,n),Angle(n))-sum(n| Combine_rl(L,n),Angle(n))))));
Constraint Flowlimit4
  IndexDomain: L;
  Definition:QFL_RN(L)=(-(sqr(sum(n|Combine_rl(L,n),Voltage(n))))*(LineSusceptance2(L)+(-
LineSusceptance(L)))+((sum(n| Combinesl(L,n),Voltage(n))*sum(n|
Combine_rl(L,n),Voltage(n)))*(Conductance(L)*(sin(sum(n| Combine_sl(L,n),Angle(n))-sum(n|
Combine_rl(L,n),Angle(n))))+((-LineSusceptance(L))*(cos(sum(n| Combine_sl(L,n),Angle(n))-sum(n|
Combine_rl(L,n),Angle(n))))))));
Constraint PLoss
  IndexDomain: L;
  Definition: PLoss(L)=Conductance(L)*(sum(n|Combine_sl(L,n),sqr(Voltage(n)))+sum(n|
Combine_rl(L,n),sqr(Voltage(n)))-2*(sum(n| Combine_sl(L,n),Voltage(n))*sum(n| Combine_rl(L,n),Voltage(n))*cos(sum(n|
Combine_sl(L,n),Angle(n))-sum(n| Combine_rl(L,n),Angle(n)))));
Constraint QLoss
  IndexDomain: L;
  Definition:QPLoss(L)=-(-LineSusceptance(L))*((sum(n|Combine_sl(L,n),sqr(Voltage(n)))+sum(n|
Combine_rl(L,n),sqr(Voltage(n))))-2*(sum(n| Combine_sl(L,n),Voltage(n))*sum(n|
Combine_rl(L,n),Voltage(n)))*(cos(sum(n| Combine_sl(L,n),Angle(n))-sum(n| Combine_rl(L,n),Angle(n)))));
Constraint Linelimit1
  IndexDomain: L;
  Definition: 0<=(sqr(RFL_SN(L))+sqr(QFL_SN(L)));
Constraint Linelimit2
  IndexDomain: L;
  Definition: (sqr(RFL_SN(L))+sqr(QFL_SN(L)))<=sqr(LineCap(L));
Constraint Linelimit3
  IndexDomain: L;
  Definition: (sqr(RFL_RN(L))+sqr(QFL_RN(L)))<=sqr(LineCap(L));

    Procedure MainExecution {
      Body: {
        !solve NetCost;
```

```
        myGMP:=GMP::Instance::Generate( NetCost );

        GMPOuterApprox::CreateStatusFile:=1;

        GMPOuterApprox::IsConvex:=1;

        GMPOuterApprox::DoOuterApproximation( myGMP );
    }
}
```

Considering the optimization results obtained from the AC STNEP model, the total cost of the system is $88.14 M with a total generation of 771.87 MW. Compared to the DC model, which has neglected the power loss owing to the assumption that reactive power and voltage magnitude are neglected, the power loss in the AC STNEP model is estimated as 11.87 MW (Tables 9.7, 9.8 and 9.9).

**Table 9.7** Real and reactive power for AC STNEP model

| Generators | Buses | $P_g$ (MW) | $Q_g$ (MW) |
|---|---|---|---|
| G1 | 1 | 100 | 65 |
| G2 | 3 | 355.56 | 41.73 |
| G3 | 6 | 316.31 | 117.41 |

**Table 9.8** Optimal power flow and line losses on the existing and prospective transmission lines for AC STNEP model

| Between buses | $F_l^+$ (MW) | $Q_l^+$ (MVAr) | $F_l^-$ (MW) | $Q_l^-$ (MVAr) | $P_{\text{loss}}$ (MW) | $Q_{\text{loss}}$ (MVAr) |
|---|---|---|---|---|---|---|
| 1 | 9.11 | 16.91 | −9.20 | −15.49 | 0.14 | 1.42 |
| 2 | 6.18 | 7.11 | −6.17 | −6.60 | 0.05 | 0.51 |
| 5 | 72.89 | 13.06 | −76.98 | −1.65 | 1.14 | 11.41 |
| 6 | 72.89 | 13.06 | −76.98 | −1.65 | 1.14 | 11.41 |
| 7 | 56.58 | 2.54 | −58.78 | 4.14 | 0.67 | 6.68 |
| 8 | 56.58 | 2.54 | −58.78 | 4.14 | 0.67 | 6.68 |
| 9 | 56.58 | 2.54 | −58.78 | 4.14 | 0.67 | 6.68 |
| 10 | 78.69 | 35.72 | −76.84 | −15.40 | 2.03 | 20.32 |
| 11 | 80.61 | 27.41 | −76.90 | −7.68 | 1.97 | 19.73 |
| 12 | 80.61 | 27.41 | −76.90 | −7.68 | 1.97 | 19.73 |
| 13 | 33.20 | 13.44 | −31.90 | −6.33 | 0.71 | 7.11 |
| 14 | 33.20 | 13.44 | −31.90 | −6.33 | 0.71 | 7.11 |

**Table 9.9** Voltage magnitude and phase angle for AC STNEP model

| Buses | Voltage (PU) | Phase angle (Rad) |
|---|---|---|
| 1 | 1.02 | 0.42 |
| 2 | 0.95 | 0.39 |
| 3 | 0.98 | 0.54 |
| 4 | 0.97 | 0.38 |
| 5 | 0.97 | 0.42 |
| 6 | 1.05 | 0.61 |

## 9.2.4 Nomenclature

### Set

$g$ existing generator
$k$ demand
$l$ transmission line

### Parameters

| | |
|---|---|
| $C_g^o$ | operation cost of the generator $g$ |
| $E_g$ | emissions cost of the generator $g$ |
| $\beta$ | duration factor |
| $P_k$ and $Q_k$ | active and reactive system load demand $k$ |
| $P_g^{\min}$ and $Q_g^{\min}$ | minimum active and reactive power capacity of the generator $g$ |
| $P_g^{\max}$ and $Q_g^{\max}$ | maximum active and reactive power capacity of the generator $g$ |
| $S_l$ | apparent power of the transmission line $l$ |
| $UB_l$ | utility budget for prospective transmission $l$ |
| $B_l$ | susceptance of transmission line |
| $V_{\min}$ and $V_{\max}$ | minimum and maximum voltage magnitude |
| $V_a$ and $V_b$ | sending and receiving end bus, respectively |
| $b_l$, $b_{lo}$ and $g_l$ | susceptance, shunt susceptance, and conductance of transmission line, respectively |

### Variables

| | |
|---|---|
| $P_g$ and $Q_g$ | active and reactive power produced by the generator $g$ |
| $P_l$ and $Q_l$ | active and reactive power flowing on transmission line $l$ |
| $P_{\text{loss}}$ and $Q_{\text{loss}}$ | active and reactive power loss on transmission line $l$ |
| $P_l^+$ and $Q_l^+$ | active and reactive power flowing into the transmission line $l$ |
| $P_l^-$ and $Q_l^-$ | active and reactive power flowing out of the transmission line $l$ |
| $z_l$ | binary variable for prospective transmission line $l$ |
| $\phi_b$ | voltage angle at bus $b$ |
| $V_b$ | voltage magnitude at bus $b$ |

## 9.3 Dynamic Transmission Network Expansion Planning (DTNEP)

Here, a DTNEP model is examined for planning system computation. This planning system approach involves making decisions about an expansion planning system at different stages of the planning horizon, making use of an integrated yearly representation of investment decisions [9]. The same objective function as in the STNEP is employed, which seeks to minimize the investment cost for new transmission lines and operational cost for the generating units in the system. This model is similar to the model in Sect. 9.2 except for the yearly representation of the time by period $t$ for the parameters and variables. Therefore, the new additional lines built over a specific period remain the same for the next period.

### 9.3.1 DC Model

The DC model for the DTNEP optimization problem can be formulated as:

$$\text{Min}\left[\frac{1}{(1+r)^t}\left(\sum_{t=1}^{n_t}\sum_{l=1}^{n_l} C_l^i z_{l,t} + \beta\sum_{t=1}^{n_t}\sum_{g=1}^{n_g} C_g^o P_{g,t}\right)\right] \tag{9.30}$$

$$C_g^o P_{g,t} = \alpha_g + \beta_g P_{g,t} + \delta_g P_{g,t}^2 \tag{9.31}$$

subject to the following constraints:

$$\sum_g P_{g,t} - \sum_{l\in sl} P_{l,t}^+ + \sum_{r\in rl} P_{l,t}^- = \sum_k P_{k,t} \quad \forall b, t \tag{9.32}$$

$$\sum_l C_l^i z_{l,t} \le UB_{l,t} \quad \forall t \tag{9.33}$$

$$z_{l,t} \in [0,1] \quad \forall l \in pl, t \tag{9.34}$$

$$P_{l,t} = B_l\left(\phi_{l\in sl,t} - \phi_{l\in rl,t}\right) \quad \forall el, t \tag{9.35}$$

$$-(1-z_{l,t})M \le P_{l,t} - B_l\left(\phi_{l\in sl,t} - \phi_{l\in rl,t}\right) \le M(1-z_{l,t}) \quad \forall pl, t \tag{9.36}$$

$$-P_l^{\max} \leq P_{l,t} \leq P_l^{\max} \quad \forall el, t \tag{9.37}$$

$$-z_{l,t} P_l^{\max} \leq P_{l,t} \leq z_{l,t} P_l^{\max} \quad \forall pl, t \tag{9.38}$$

$$P_g^{\min} \leq P_{g,t} \leq P_g^{\max} \quad \forall g, t \tag{9.39}$$

$$-\pi \leq \phi_{b,t} \leq \pi \quad \forall b, t. \tag{9.40}$$

The desired variable results to be obtained by the optimization model are $z_{l,t}$, $\phi_{b,t}$, $P_{g,t}$ and $P_{l,t}$.

The DC model of the dynamic TNEP problem developed was implemented using the IEEE Garver six-bus test system as given in Fig. 9.1. When considering the dynamic planning approach, expansion planning will be considered for a certain number of planning periods to ascertain the optimal planning results for the dynamic planning system [10]. In this chapter, data for the load on the six-bus test system have been modified and can be found in Table 9.10. The planning horizon is targeted to be 10 years with annual load growth of 1.1% at the planning period and the discount rate is 10%.

The model constraints in Eqs. (9.32)–(9.40) are the same as what was obtained in Eqs. (9.3)–(9.11). The only difference between these constraints is the yearly representation of the time by period $t$ for the parameters and variables.

The AIMMS code for solving the example for a DC model of dynamic TNEP is given below:

**Table 9.10** Modified load data for 10-year planning horizon

| Load bus location | Period (year) | | | | | | | | | |
|---|---|---|---|---|---|---|---|---|---|---|
| | 1 | 2 | 3 | 4 | 5 | 6 | 7 | 8 | 9 | 10 |
| 1 | 40 | 44 | 48 | 52 | 56 | 60 | 64 | 68 | 72 | 80 |
| 2 | 120 | 132 | 144 | 156 | 168 | 180 | 192 | 204 | 216 | 240 |
| 3 | 20 | 22 | 24 | 26 | 28 | 30 | 32 | 34 | 36 | 40 |
| 4 | 80 | 88 | 96 | 104 | 112 | 120 | 128 | 136 | 144 | 160 |
| 5 | 120 | 132 | 144 | 156 | 168 | 180 | 192 | 204 | 216 | 240 |

```
Model Main_DC-DTNEP
  Set generator
    Index: g;
  Set nodes
    Index: n;
  Set demand
    Index: k;
  Set periods
    Index: t;
  Set line
    Index: l;
  Set existinglines
    SubsetOf: line;
  Set prospectivelines
    SubsetOf: line;
Parameters
  Parameter CombineGen
    IndexDomain: (g,n);
  Parameter Combinedemand
    IndexDomain: (k,n);
  Parameter Combine_sl
    IndexDomain: (l,n);
  Parameter Combine_rl
    IndexDomain: (l,n);
  Parameter demands
    IndexDomain: (k,t);
  Parameter fmax
    IndexDomain: l;
  Parameter linesusceptance
    IndexDomain: l;
  Parameter Pgmax
    IndexDomain: g;
  Parameter Costa
    IndexDomain: g;
  Parameter Costb
    IndexDomain: g;
  Parameter Costc
    IndexDomain: g;
  Parameter Linvestcost
    IndexDomain: l;
  Parameter M;
  Parameter duration;
  Parameter d;
  Parameter dd
    IndexDomain: t;
    Definition: (1/(1+0.085)^v(t));
  Parameter v
    IndexDomain: t;
Variables
  Variable ZL
    IndexDomain: (l,t);
    Range: binary;
  Variable Pg
    IndexDomain: (g,t);
    Range: nonnegative;
  Variable flow
    IndexDomain: (l,t);
    Range: nonnegative;
  Variable Angle
    IndexDomain: (n,t);
    Range: free;
Objective Function
  Variable Totalcost
    Range: free;
    Definition: sum(t,(1/(1+d)^v(t))*(sum(l|prospectivelines, Linvestcost(l)*ZL(l,t))+(duration*sum(g,
(Costa(g)+(Costb(g)*Pg(g,t))+(Costc(g)*Pg(g,t)^2))))));
  MathematicalProgram Leastcost {
```

```
    Objective: Totalcost;
    Direction: minimize;
    Constraints: AllConstraints;
    Variables: AllVariables;
    Type: Automatic;
Constraints
  Constraint Powerbalance
    IndexDomain: (n,t);
    Definition: sum(g|MapGen(g,n),Pg(g,t))-sum(l| Combine_sl(l,n),flow(l,t))+sum(l| Combine_rl(l,n),flow(l,t))=sum(k|
Combine_demand(k,n),demands(k,t));
  Constraint linelimit1
    IndexDomain: (l,t)|ord(l)<=6;
    Definition: -fmax(l)<=flow(l,t);
  Constraint linelimit2
    IndexDomain: (l,t)|ord(l)<=6;
    Definition: flow(l,t)<=fmax(l);
  Constraint linelimit3
    IndexDomain: (l,t)|ord(l)>6;
    Definition: -z(l,t)*fmax(l)<=flow(l,t);
  Constraint linelimit4
    IndexDomain: (l,t)|ord(l)>6;
    Definition: flow(l,t)<=z(l,t)*fmax(l);
  Constraint genlimit1
    IndexDomain: (g,t);
    Definition: 0<=Pg(g,t);
  Constraint genlimit2 {
    IndexDomain: (g,t);
    Definition: Pg(g,t)<=Pgmax(g);
  Constraint buslimit
    IndexDomain: (n,t);
    Definition: -3.14<=ang(n,t)<=3.14;
  Constraint flowlimit1
    IndexDomain: (l,t)|ord(l)<=6;
    Definition: flow(l,t)=lnesusceptance(l)*(sum(n| Combine_sl(l,n),Angle(n,t))-sum(n| Combine_rl(l,n),Angle(n,t)));
  Constraint flowlimit2
    IndexDomain: (l,t)|ord(l)>6;
    Definition:          (-(1-ZL(l,t))*M)<=linesusceptance(l)*(sum(n|          Combine_sl(l,n),Angle(n,t))-sum(n|
Combine_rl(l,n),Angle(n,t)));
  Procedure MainExecution {
    Body: {
      solve Leastcost;
    }
  }
```

The simulation results were considered for the dynamic planning approach to analyze the performance of the proposed model. Tables 9.11 and 9.12 present the year-to-year optimal power generated by the generating units and the OPF in the existing and prospective transmission lines. From the results, it can be seen that generators are utilized based on their annual demand, contrary to a single-stage planning approach as in Sect. 9.2.1, where the planning is done at once and allows growth in demand, leaving the generators unused. Secondly, most of the transmission lines were operating at their full capacity when a dynamic approach was considered against the static model in Sect. 9.2.1.

**Table 9.11** Optimal generated power for DC DTNEP model

| Generators | Bus | Period (year) | | | | | | | | | |
|---|---|---|---|---|---|---|---|---|---|---|---|
| | | 1 | 2 | 3 | 4 | 5 | 6 | 7 | 8 | 9 | 10 |
| 1 | 1 | 125 | 144 | 150 | 132 | 150 | 120 | 139 | 150 | 127 | 150 |
| 2 | 3 | 155 | 174 | 206 | 162 | 182 | 150 | 169 | 196 | 157 | 210 |
| 3 | 6 | 100 | 100 | 100 | 200 | 200 | 300 | 300 | 300 | 400 | 400 |

**Table 9.12** Optimal power flow on the transmission lines for DC DTNEP model

| Lines | Period (year) | | | | | | | | | |
|---|---|---|---|---|---|---|---|---|---|---|
| | 1 | 2 | 3 | 4 | 5 | 6 | 7 | 8 | 9 | 10 |
| 1–2 | 14.39 | 20.17 | 21.66 | 37.22 | 42.67 | 10.32 | 16.10 | 19.24 | 0.74 | 7.36 |
| 1–4 | 22.86 | 26.96 | 29.27 | 37.49 | 41.46 | 29.55 | 33.66 | 36.66 | 9.56 | 15.70 |
| 1–5 | 47.76 | 52.87 | 51.06 | 5.29 | 9.87 | 20.13 | 25.24 | 26.10 | 44.70 | 46.94 |
| 2–3 | 57.14 | 61.04 | 66.73 | 66.51 | 70.54 | 90.45 | 94.34 | 99.34 | 34.44 | 44.30 |
| 2–4 | 62.76 | 72.87 | 89.06 | 85.29 | 95.87 | 60.13 | 70.24 | 84.10 | 49.70 | 76.94 |
| 3–5 | 72.24 | 79.13 | 92.94 | 50.71 | 58.13 | 59.87 | 66.76 | 77.90 | 71.30 | 93.06 |
| 2–6 | 100 | 100 | 100 | 100 | 100 | 100 | 100 | 100 | 100 | 100 |
| 2–6 | – | – | – | – | – | 100 | 100 | 100 | 100 | 100 |
| 4–6 | – | – | – | – | – | – | – | – | 100 | 100 |
| 5–6 | – | – | – | 100 | 100 | 100 | 100 | 100 | 100 | 100 |

## *9.3.2 Nomenclature*

### Set

$g$ existing generator
$k$ demand
$t$ time

### Parameters

$C_g^o$ operation cost of the generator $g$
$E_g$ emissions cost of the generator $g$
$\beta$ duration factor
$P_{k,t}$ system load demand $k$ at time $t$
$P_g^{\min}$ and $P_g^{\max}$ minimum and maximum power capacity of the generator $g$
$UB_{l,t}$ utility budget for prospective transmission line $l$ at time $t$
$B_l$ susceptance of transmission line

**Variables**

| | |
|---|---|
| $P_{g,t}$ | power produced by the generator $g$ at time $t$ |
| $P_{l,t}$ | power flowing on transmission line $l$ at time $t$ |
| $P_{l,t}^{+}$ and $P_{l,t}^{-}$ | power flowing into and out of the transmission line $l$ at time $t$ |
| $z_{l,t}$ | binary variable for prospective transmission line $l$ at time $t$ |
| $\phi_{b,t}$ | voltage angle at bus $b$ at time $t$ |

## 9.4 Demand Response Integrated Dynamic Transmission Network Expansion Planning

This section provides information on the system scheduling between the demand side and the transmission side by the system operator. Adequate consideration of the demand side with the TNEP is very important to achieve an optimal expansion planning strategy aimed at smart electricity [11]. The DR is seen as a means of reshaping the system load profile to achieve improved system efficiency [12, 13]. DR can be defined as changes that occur in electricity usage by demand-side consumers from the normal consumption patterns with respect to changes in electricity price over time or financial incentives offered to consumers to reduce electricity usage at times of high wholesale market prices to obtain higher energy efficiency and improve system stability [14]. The peak loads mostly occur over a restricted period; therefore, it is logical to consider the DR aggregators as an alternative means to meet the peak load as well as a way of reducing TNEP investment costs.

The mathematical model for the dynamic TNEP optimization problem with DR can be formulated as:

$$\text{Min}\left[\frac{1}{(1+r)^t}\left(\sum_{t=1}^{n_t}\sum_{l=1}^{n_l} C_l^i z_{l,t} + \beta\sum_{t=1}^{n_t}\sum_{g=1}^{n_g} C_g^o P_{g,t} + \lambda\sum_{t=1}^{n_t}\sum_{y}^{n_y}\delta_y\left(C_y^r \alpha_{y,t}^r\right)\right)\right] \tag{9.41}$$

$$C_g^o P_{g,t} = \alpha_g + \beta_g P_{g,t} + \delta_g P_{g,t}^2 \tag{9.42}$$

$$\sum_{g} P_{g,t} - \sum_{l\in sl} P_{l,t}^{+} + \sum_{r\in rl} P_{l,t}^{-} + \sum_{y} \alpha_{y,t}^r = \sum_{k} P_{k,t} \quad \forall b,t \tag{9.43}$$

$$0 \le \alpha_{y,t}^r \le \overline{\alpha^r} \tag{9.44}$$

$$\overline{\alpha^r} \leq P_{k,t} \tag{9.45}$$

$$\sum_l C_l^i z_{l,t} \leq UB_{l,t} \quad \forall t \tag{9.46}$$

$$z_{l,t} \in [0,1] \quad \forall l \in pl, t \tag{9.47}$$

$$P_{l,t} = B_l(\phi_{l \in sl,t} - \phi_{l \in rl,t}) \quad \forall el, t \tag{9.48}$$

$$-(1 - z_{l,t})M \leq P_{l,t} - B_l(\phi_{l \in sl,t} - \phi_{l \in rl,t}) \leq M(1 - z_{l,t}) \quad \forall pl, t \tag{9.49}$$

$$-P_l^{\max} \leq P_{l,t} \leq P_l^{\max} \quad \forall el, t \tag{9.50}$$

$$-z_{l,t}P_l^{\max} \leq P_{l,t} \leq z_{l,t}P_l^{\max} \quad \forall pl, t \tag{9.51}$$

$$P_g^{\min} \leq P_{g,t} \leq P_g^{\max} \quad \forall g, t \tag{9.52}$$

$$-\pi \leq \phi_{b,t} \leq \pi \quad \forall b, t. \tag{9.53}$$

Equation (9.41) is the objective function, which seeks to minimize the investment cost for new transmission lines, operational cost for the generators, and the DR costs in the system. Equation (9.42) defines the quadratic cost function of the operational cost as a function of the power output. In constraint (9.43), the DR has been integrated into the power balance constraint and this gives the total power generated by the generators, the power flowing out of and into the lines and the optimal DR power out, which must equal the load demand [9]. Equations (9.44) and (9.45) are the DR constraints, which explain that the DR maximum limit must not be exceeded at any given time and that the maximum boundary of the DR must be within the range of the maximum load limit at any time. Constraints (9.46)–(9.53) are the same as constraints (9.33)–(9.40).

The available time slot for DR on the network is taken as between two and four hours and the average DR was obtained from the percentage of available time with the load at each bus. The DR cost is taken to be \$105/MWh as obtained in [15].

The AIMMS code for solving the example for a mathematical model of dynamic TNEP with DR is given below:

```
Model Main_DTNEP-DR
  Set generator
    Index: g;
  Set nodes
    Index: n;
  Set demand
    Index: k;
  Set periods
    Index: t;
  Set line
    Index: l;
  Set existinglines
    SubsetOf: line;
  Set prospectivelines
    SubsetOf: line;
  Set DR
    Index: dr;
  Parameter CombineDR
    IndexDomain: (dr,n);
  Parameter UpperDR
    IndexDomain: (dr,t);
  Parameter CombineGen
    IndexDomain: (g,n);
  Parameter Combinedemand
    IndexDomain: (k,n);
  Parameter Combine_sl
    IndexDomain: (l,n);
  Parameter Combine_rl
    IndexDomain: (l,n);
  Parameter demands
    IndexDomain: (k,t);
  Parameter fmax
    IndexDomain: l;
  Parameter linesusceptance
    IndexDomain: l;
  Parameter Pgmax
    IndexDomain: g;
  Parameter Costa
    IndexDomain: g;
  Parameter Costb
    IndexDomain: g;
  Parameter Costc
    IndexDomain: g;
  Parameter Linvestcost
    IndexDomain: l;
  Parameter M;
  Parameter duration;
  Parameter d;
  Parameter dd
    IndexDomain: t;
    Definition: (1/(1+0.085)^v(t));
  Parameter v
    IndexDomain: t;
  Parameter duration1;
  Parameter duration2;
  Parameter DRduration
    IndexDomain: dr;
  Parameter CostDR
      IndexDomain: dr;
  Variables
    Variable DRR
      IndexDomain: (dr,t);
      Range: nonnegative;
    Variable ZL
      IndexDomain: (l,t);
      Range: binary;
    Variable Pg
```

```
      IndexDomain: (g,t);
      Range: nonnegative;
   Variable flow
      IndexDomain: (l,t);
      Range: nonnegative;
   Variable Angle
      IndexDomain: (n,t);
      Range: free;
Objective function
   Variable Totalcost
      Range: free;
      Definition: sum(t,(1/(1+d)^v(t))*(sum(l|prospectivelines, Linvestcost(l)*ZL(l,t))+(duration1*sum(g,
(Costa(g)+(Costb(g)*Pg(g,t))+(Costc(g)*Pg(g,t)^2))))+(sum(dr,(duration2*DRduration(dr)*CostDR(dr)*DRR(dr,t))))));
   MathematicalProgram Leastcost {
      Objective: Totalcost;
      Direction: minimize;
      Constraints: AllConstraints;
      Variables: AllVariables;
      Type: Automatic;
Constraints
   Constraint DRRlimit1
      IndexDomain: (dr,t);
      Definition: 0<=DRR(dr,t);
   Constraint DRRlimit2 {
      IndexDomain: (dr,t);
      Definition: DRR(dr,t)<=UpperDR(dr,t);
   Constraint powerbalance
      IndexDomain: (n,t);
      Definition: sum(g| CombineGen(g,n),Pg(g,t))-sum(l| Combine_sl(l,n),flow(l,t))+sum(l|
Combine_rl(l,n),flow(l,t))+sum(dr| CombineDR(dr,n),DRR(dr,t))=sum(k| Combinedemand(k,n),demands(k,t));
   Constraint linelimit1
      IndexDomain: (l,t)|ord(l)<=6;
      Definition: -fmax(l)<=flow(l,t);
   Constraint linelimit2
      IndexDomain: (l,t)|ord(l)<=6;
      Definition: flow(l,t)<=fmax(l);
   Constraint linelimit3
      IndexDomain: (l,t)|ord(l)>6;
      Definition: -z(l,t)*fmax(l)<=flow(l,t);
   Constraint linelimit4
      IndexDomain: (l,t)|ord(l)>6;
      Definition: flow(l,t)<=z(l,t)*fmax(l);
   Constraint genlimit1
      IndexDomain: (g,t);
      Definition: 0<=Pg(g,t);
   Constraint genlimit2
      IndexDomain: (g,t);
      Definition: Pg(g,t)<=Pgmax(g);
   Constraint buslimit
      IndexDomain: (n,t);
      Definition: -3.14<=Angle(n,t)<=3.14;
   Constraint flowlimit1
      IndexDomain: (l,t)|ord(l)<=6;
      Definition: flow(l,t)=linesusceptance(l)*(sum(n| Combine_sl(l,n),Angle(n,t))-sum(n| Combine_rl(l,n),Angle(n,t)));
   Constraint flowlimit2
      IndexDomain: (l,t)|ord(l)>6;
      Definition: (-(1-ZL(l,t))*M)<=linesusceptance(l)*(sum(n| Combine_sl(l,n),Angle(n,t))-sum(n|
Combine_rl(l,n),Angle(n,t)));
   Procedure MainExecution {
      Body: {
         solve Leastcost;
      }
   }
```

The DR is integrated into the dynamic TNEP in this section and the optimization results obtained are examined. Tables 9.13 and 9.14 give the year-to-year optimal power generated by the generating units and the OPF in the existing and prospective transmission lines. Table 9.15 presents the optimal load curtailment owing to the presence of 100% DR in the network. In this study, we introduced a 100% penetration level of DR into the network. A different level of penetration of DR can be examined to determine the performance of DR in the TNEP process. It can be observed from the results obtained that the total cost for the system planning is $667 M, as against $730 M obtained in DTNEP (Sect. 9.3.1). Therefore, integration of DR in the network serves as a trade-off for TNEP, thus minimizing the investment cost of new transmission lines in the system.

## 9.5 Summary

This chapter provides an analysis of the TNEP problem with different planning approaches without and with integration of DR in the power system. The solutions for the TNEP problems for both DC and AC planning methods are provided. These solutions aim at aiding the energy trading between the utilities and the end users. Different examples are illustrated in the chapter with an appropriate AIMMS code.

## 9.6 Nomenclature

The nomenclature used in this chapter is given below:

**Set**

| | |
|---|---|
| $g$ | generator |
| $k$ | demand |
| $l$ | transmission line |
| $el$ | existing transmission line |
| $pl$ | prospective transmission line |
| $r$ | demand response |
| $t$ | time |
| $y$ | source of demand response |

**Parameters**

| | |
|---|---|
| $C_g^o$ | operational cost of the generator $g$ |
| $E_g$ | emissions cost of the generator $g$ |
| $\beta$ | duration factor |
| $P_{k,t}$ | system load demand $k$ at time $t$ |
| $P_g^{\min}$ and $P_g^{\max}$ | minimum and maximum power capacity of the generator $g$ |
| $\delta_y$ | available time slot at demand response source $y$ |
| $C_{y,t}^r$ | cost of power curtailed by demand response $r$ at source $y$ in time $t$ |
| $\overline{\alpha^r}$ | maximum capacity value of demand response $r$ |

**Table 9.13** Optimal generated power for DTNEP-DR model

| Generators | Bus | Period (year) | | | | | | | | | |
|---|---|---|---|---|---|---|---|---|---|---|---|
| | | 1 | 2 | 3 | 4 | 5 | 6 | 7 | 8 | 9 | 10 |
| 1 | 1 | 150 | 141.25 | 150 | 150 | 150 | 150 | 150 | 150 | 150 | 150 |
| 2 | 3 | 197.59 | 171.25 | 200 | 223.83 | 223.40 | 204.61 | 207.62 | 187.50 | 217.86 | 236.67 |
| 3 | 6 | – | 100 | 100 | 100 | 100 | 200 | 200 | 300 | 300 | 300 |

**Table 9.14** Optimal power flow on the lines for DTNEP-DR model

| Lines | Period (year) | | | | | | | | | |
|---|---|---|---|---|---|---|---|---|---|---|
| | 1 | 2 | 3 | 4 | 5 | 6 | 7 | 8 | 9 | 10 |
| 1 | 41.11 | 20.47 | 22.98 | 23.53 | 21.23 | 13.16 | 12.29 | 21.11 | 30.71 | 14.18 |
| 2 | 31.48 | 27.09 | 29.83 | 31.74 | 30.44 | 9.73 | 11.05 | 37.45 | 22.15 | 34.14 |
| 3 | 40.74 | 53.35 | 53.19 | 47.06 | 47 | 72.11 | 68 | 29.11 | 31.14 | 28.35 |
| 4 | 36.11 | 60.91 | 66.17 | 72.26 | 71.96 | 10.27 | 16.95 | 98.55 | 21.85 | 100 |
| 5 | 100 | 72.44 | 87.19 | 100 | 97.73 | 77.11 | 78.29 | 81.44 | 100 | 100 |
| 6 | 79.26 | 78.65 | 90.81 | 100 | 100 | 100 | 100 | 74.89 | 84.86 | 100 |
| 7 | – | – | 100 | 100 | 100 | 100 | 100 | 100 | 100 | 100 |
| 8 | – | – | – | – | – | – | 100 | 100 | 100 | 100 |
| 9 | – | – | – | – | – | – | – | 100 | 100 | 100 |
| 10 | – | – | – | – | – | – | – | 100 | 100 | 100 |

**Table 9.15** Optimal load curtailment for 100% participating factor for DTNEP-DR model

| Demand response | Available time slot, h | Period (year) | | | | | | | | | |
|---|---|---|---|---|---|---|---|---|---|---|---|
| | | 1 | 2 | 3 | 4 | 5 | 6 | 7 | 8 | 9 | 10 |
| 1 | 2 | 3.33 | 3.67 | 4 | 4.33 | 4.67 | 5 | 5.33 | 5.67 | 6 | 6.67 |
| 2 | 3 | 15 | – | – | 4.73 | 21 | – | 18.38 | – | 7.14 | 25.82 |
| 3 | 2 | 1.67 | 1.83 | 2 | 2.17 | 2.33 | 2.5 | 2.67 | 2.83 | 3 | 3.33 |
| 4 | 4 | 12.41 | – | – | – | 9.60 | – | – | – | – | 25.86 |
| 5 | 3 | – | – | – | 8.94 | 21 | 7.89 | 24 | – | – | 11.65 |

| | |
|---|---|
| $UB_{l,t}$ | utility budget for prospective transmission line $l$ at time $t$ |
| $B_l$ | susceptance of transmission line |
| $M$ | disjunctive factor |
| $r$ | discount factor |

## Variables

| | |
|---|---|
| $P_{g,t}$ | power produced by the generator $g$ at time $t$ |
| $P_{l,t}$ | power flowing on transmission line $l$ at time $t$ |
| $P^{+}_{l,t}$ and $P^{-}_{l,t}$ | power flowing into and out of the transmission line $l$ at time $t$ |
| $\alpha^{r}_{y,t}$ | active power curtailed by demand response $r$ at source $y$ in time $t$ |
| $z_{l,t}$ | binary variable for prospective transmission line $l$ at time $t$ |
| $\phi_{b,t}$ | voltage angle at bus $b$ at time $t$ |

## References


1. Garc R, Roberto M (2017) Dynamic robust transmission expansion planning. IEEE Trans Power Syst 32(4):2618–2628
2. Gbadamosi SL, Nwulu NI, Sun Y (2018a) Multi-objective optimisation for composite generation and transmission expansion planning considering offshore wind power and feed-in tariffs. IET Renew Power Gener 12(14):1687–1697
3. Gbadamosi SL, Nwulu, NI, Sun Y (2018) Optimal scheduling in offshore wind farms considering harmonic losses and locational marginal pricing. In: 2018 IEEE PES-IAS 5th power Africa conference, pp 658–663
4. Gbadamosi SL, Nwulu NI (2020) Reliability assessment of composite generation and transmission expansion planning incorporating renewable energy sources. J Renew Sustain Energy 026301(12)
5. Gbadamosi SL, Nwulu NI (2020) Optimal power dispatch and reliability analysis of hybrid CHP-PV-wind systems in farming applications. Sustainability 12
6. Gbadamosi SL, Nwulu NI, Damisa U (2019) Impact of power losses on optimal expansion planning using quadratic programming. In: 6th IEEE international conference on engineering technologies and applied sciences (ICETAS), pp 1–4
7. Zhang H (2013) Transmission expansion planning for large power systems. Arizona State University
8. Gbadamosi SL, Nwulu NI, Sun Y (2018b) Harmonic and power loss minimization in power systems incorporating renewable energy sources and locational marginal pricing. J Renew Sustain Energy 10:1–24
9. Gbadamosi SL, Nwulu NI (2020) A multi-period composite generation and transmission expansion planning model incorporating renewable energy sources and demand response. Sustain Energy Technol Assess 39:100726
10. Gbadamosi SL, Nwulu NI (2019) A dynamic optimal expansion planning model integrating renewable energy sources. In: 11th international conference on applied energy, pp 150–155
11. Zhang Q, Mclellan BC, Tezuka T, Ishihara KN (2013) An integrated model for long-term power generation planning toward future smart electricity systems. Appl Energy 112:1424–1437
12. Nwulu NI, Xia X (2017) Optimal dispatch for a microgrid incorporating renewables and demand response. Renew Energy 101:16–28
13. Nwulu NI, Xia X (2015a) Implementing a model predictive control strategy on the dynamic economic emission dispatch problem with game theory based demand response programs. Energy 91:404–419
14. Nwulu NI, Xia X (2015b) Multi-objective dynamic economic emission dispatch of electric power generation integrated with game theory based demand response programs. Energy Convers Manag 89:963–974
15. Zhang N, Hu Z, Springer C, Li Y, Shen B (2016) A bi-level integrated generation-transmission planning model incorporating the impacts of demand response by operation simulation. Energy Convers Manag 123:84–94

# Chapter 10
# Generation Capacity Expansion Planning

## 10.1 Introduction

This chapter provides an analysis of GCEP. GCEP is an investment planning system aimed at guaranteeing minimal costs and ensuring customer satisfaction [2]. The main objective is to expand the power system network with the incorporation of renewable energy-generating units to serve the growing energy demand. GCEP determines the location, time, technology, and optimal capacity of prospective generators to be built for system operation [3]. Here, a centralized approach is considered for determining the expansion planning that is most suitable for system operations. The information provided will encourage private investors to construct electricity-generating facilities based on the determined expansion plan that will give optimal operation of the power system with a reliable and efficient power supply. Here, we describe the GCEP problem models based on real and reactive power capacity expansion planning systems, involving static and dynamic models.

## 10.2 Load Forecasting

Load forecasting in an electric power system plays an important role in power system investment, expansion planning, and operations. The decisions obtained from the forecasting of demand will ensure stable, reliable, and economic operation of the power system [4]. Load forecasting is related to power grid planning; therefore, it can be divided into long-term, medium-term, and short-term load forecasting. To obtain an optimal planning system as well as possible solutions, power system planners must ensure that their models and analytical approaches consider an increase in electricity usage and customer flexibility that normally alters energy demand and the load profile.

 
N. Nwulu and S. L. Gbadamosi, *Optimal Operation and Control of Power Systems Using an Algebraic Modelling Language*, Green Energy and Technology,
https://doi.org/10.1007/978-3-030-00395-1_10

## 10.3 Real Power Capacity Expansion Planning

Here, we consider the active power of the GCEP problem; therefore, the reactive power is neglected. Hence, this model is considered a DC model because it is similar to the DC model assumption. GCEP is also regarded as a long-term planning horizon because these plans are capital-intensive and must ensure that demands will be met even in worst-case scenarios [5]. As discussed in Chap. 9, GCEP has two expansion planning strategies, namely static and dynamic models. Here, we provide solutions for the static and dynamic GCEP problems for a DC model in AIMMS.

### *10.3.1 Static Model*

In this section, a static planning approach is considered for GCEP models and expansion planning is determined for a short-time planning horizon. The mathematical formulation presented below seeks to minimize the investment cost of prospective renewable energy-generating units, operational cost, and emissions for the existing fossil fuel-generating units in the system.

The DC model for the optimization problem can be formulated as:

$$Min\left[\sum_{\Re=1}^{n_{\Re}} C_{\Re}^{i} P_{\Re}^{\max} + \beta \sum_{g=1}^{n_g} C_{\mathrm{g}}^{\mathrm{o}} P_{\mathrm{g}} + \beta \sum_{g=1}^{n_g} E_{\mathrm{g}} P_{\mathrm{g}}\right] \tag{10.1}$$

$$C_{\mathrm{g}}^{\mathrm{o}} P_{\mathrm{g}} = \alpha_{\mathrm{g}} + \beta_{\mathrm{g}} P_{\mathrm{g}} + \delta_{\mathrm{g}} P_{\mathrm{g}}^{2} \tag{10.2}$$

$$E_{\mathrm{g}} P_{\mathrm{g}} = a_{\mathrm{g}} + b_{\mathrm{g}} P_{\mathrm{g}} + c_{\mathrm{g}} P_{\mathrm{g}}^{2} \tag{10.3}$$

subject to the following constraints:

$$\sum_{\mathrm{g}} P_{\mathrm{g}} + \sum_{\Re} P_{\Re} - \sum_{\mathrm{l}\in\mathrm{sl}} P_{\mathrm{l}}^{+} + \sum_{\mathrm{r}\in\mathrm{rl}} P_{\mathrm{l}}^{-} = \sum_{\mathrm{k}} P_{\mathrm{k}} \forall b \tag{10.4}$$

$$\sum_{\Re} C_{\Re}^{\mathrm{i}} P_{\Re}^{\max} \le UB_{\Re} \tag{10.5}$$

$$0 \le P_{\Re}^{\max} \le \overline{P_{\Re}} \forall \Re \tag{10.6}$$

$$P_{\Re}^{\max} = \sum_{\mathrm{c}} \alpha_{\Re c} P_{\Re c} \forall \Re \tag{10.7}$$

$$\sum_{\mathrm{c}} \alpha_{\Re \mathrm{c}} = 1 \forall \Re \tag{10.8}$$

$$P_l = B_l(\phi_{l\in sl} - \phi_{l\in rl})\forall l \tag{10.9}$$

$$-P_l^{max} \leq P_l \leq P_l^{max}\forall l \tag{10.10}$$

$$0 \leq P_{\Re} \leq P_{\Re}^{max}\forall \Re \tag{10.11}$$

$$P_g^{min} \leq P_g \leq P_g^{max}\forall g \tag{10.12}$$

$$-\pi \leq \phi_b \leq \pi\forall b. \tag{10.13}$$

The desired variable results to be obtained by the optimization model are $P_{\Re}$, $P_{\Re}^{max}$, $\phi_b$, $P_g$, and $P_l$.

The objective functions in Eq. (10.1) are categorized into two, namely:

(a) $\sum_{\Re=1}^{n_{\Re}} C_{\Re}^{i} P_{\Re}^{max}$ represents the annualized cost of constructing new renewable energy-generating units.

(b) $\sum_{g=1}^{n_g} C_g^{o} P_g$ and $\sum_{g=1}^{n_g} E_g P_g$ represent the cost and emission of operating the generating units.

The investment cost for the prospective renewable energy-generating units is annualized; therefore, a duration factor (8760 h) [6] is multiplied with operational cost and emission of the fossil fuel-generating units to make it comparable with the first objective function.

For simplicity purposes, a quadratic cost function is applied for the fuel cost and emission functions of the fossil fuel-generating units. In Eqs. (10.2) and (10.3), the fuel cost in \$/h and Ib/h is given as a function of its output power [7].

Constraint (10.4) ensures that at any bus b, the capacity sum of power from fossil fuel-generating units, renewable energy-generating units, and the power flowing in and out of the transmission lines equals the load demand [8]. Equation (10.5) is the constraint for the investment budget for the prospective renewable energy-generating units to be built and it states that the cost of constructing a new renewable energy-generating unit must not exceed the available funds. Constraints (10.6) to (10.8) are the constraints for the prospective renewable energy-generating units and ensure that the optimal capacity limit for the prospective renewable energy-generating units does not exceed the upper limit. Constraint (10.9) defines the power flow on the transmission lines constraint and Constraint (10.10) ensures that the flow limit does not exceed its maximum limit. Constraints (10.11) and (10.12) are the generation limits for prospective renewable energy-generating units and fossil fuel-generating units and ensure that the optimal power from prospective renewable energy- and fossil fuel-generating units does not exceed the stipulated boundary. Constraint (10.13) ensures that the phase angle at each bus is kept within a safe limit.

Figure 10.1 depicts the modified IEEE Garver six-bus test system and the proposed renewable energy-generating unit locations on the system network. The data for the modified system are also presented in Table 10.1; it consists of three fossil fuel generators and three RES generators, five loads, and ten transmission lines. The data for the transmission lines are depicted in Table 10.2 and all the details of the modified six-bus test system can be found in [3].

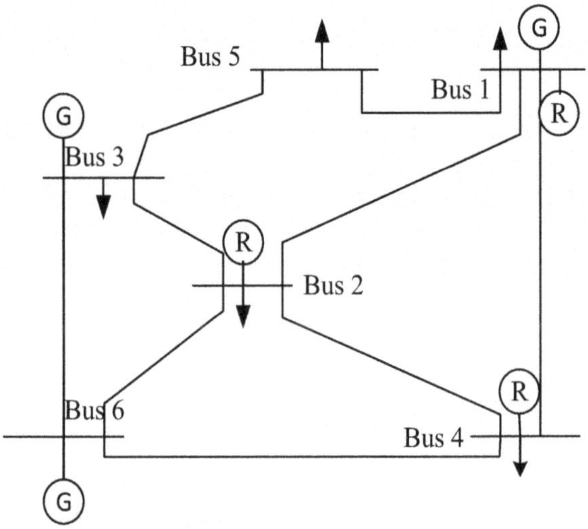


**Fig. 10.1** Modified IEEE Garver six-bus system network [3]

**Table 10.1** Modified data for IEEE Garver six-bus system example

| Nodes | Generator parameters | | | | Load parameters | | Renewable energy parameters | | | |
|---|---|---|---|---|---|---|---|---|---|---|
| | $P_g^{min}$ (MW) | $P_g^{max}$ (MW) | $Q_g^{min}$ (MVAr) | $Q_g^{max}$ (MVAr) | $P_d$ (MW) | $Q_d$ (MVAr) | $P_R^{min}$ (MW) | $P_R^{max}$ (MW) | $Q_R^{min}$ (MVAr) | $Q_R^{max}$ (MVAr) |
| 1 | 0 | 160 | −10 | 65 | 160 | 32 | 0 | 400 | −85 | 150 |
| 2 | – | – | – | – | 480 | 96 | 0 | 400 | −85 | 150 |
| 3 | 0 | 360 | −10 | 150 | 80 | 16 | 0 | – | – | – |
| 4 | – | – | – | – | 320 | 64 | 0 | 500 | −100 | 160 |
| 5 | – | – | – | – | 300 | 75 | – | – | – | – |
| 6 | 0 | 610 | −10 | 200 | – | – | – | – | – | – |

**Table 10.2** Parameters for transmission lines

| Between buses | Line capacity (MW) | Susceptance, $B_L$ |
|---|---|---|
| 1–2 | 100 | 250 |
| 1–4 | 80 | 133 |
| 1–5 | 100 | 500 |
| 2–3 | 100 | 250 |
| 2–4 | 100 | 500 |
| 3–5 | 100 | 250 |
| 2–6 | 100 | 250 |
| 3–5 | 100 | 333 |
| 4–6 | 100 | 500 |

The AIMMS code for solving the static DC GCEP model is given below:

```
Model Main_FinalSixbus {
  Set Nodes {
    Index: n;
  }
  Set EGen {
    Index: g;
  }
  Set CGen {
    Index: c;
  }
  Set Lines {
    Index: L;
  }
  Set ELines {
    SubsetOf: Lines;
  }
  Set CLines {
    SubsetOf: Lines;
  }
  Set Dem {
    Index: d;
  }
  Set COpt {
    Index: q;
  }
  Parameter MapEGen {
    IndexDomain: (g,n);
  }
  Parameter MapCGen {
    IndexDomain: (c,n);
  }
  Parameter MapDem {
    IndexDomain: (d,n);
  }
  Parameter MapSL {
    IndexDomain: (L,n);
  }
  Parameter MapRL {
    IndexDomain: (L,n);
  }
  Parameter LineCap {
    IndexDomain: (L);
  }
  Parameter LineSusc {
    IndexDomain: (L);
  }
  Parameter CGenCap {
    IndexDomain: (c,q);
  }
  Parameter EGenCap {
    IndexDomain: (g);
  }
  Parameter UPBCC {
    IndexDomain: (c);
  }
  Parameter ECosta {
    IndexDomain: (g);
```

```
}
Parameter ECostb {
  IndexDomain: (g);
}
Parameter ECostc {
  IndexDomain: (g);
}
Parameter EEa {
  IndexDomain: (g);
}
Parameter EEb {
  IndexDomain: (g);
}
Parameter EEc {
  IndexDomain: (g);
}
Parameter Load {
  IndexDomain: (d);
}
Parameter LInvstCost {
  IndexDomain: (L);
}
Parameter CInvstCost {
  IndexDomain: (c);
}
Parameter LBdgt;
Parameter CBdgt;
Parameter BigM;
Parameter Dur;
Parameter w1;
Parameter w2;
Parameter w3;
Parameter w4;
Parameter w5;
Parameter FIT {
  Definition: 0.3*sum(c,CInvstCost(c)*MaxPCC(c));
}
Variable LB {
  IndexDomain: (L);
  Range: binary;
}
Variable CB {
  IndexDomain: (c,q);
  Range: binary;
}
Variable EGenOP {
  IndexDomain: (g);
  Range: nonnegative;
}
Variable CGenOP {
  IndexDomain: (c);
  Range: nonnegative;
}
Variable Flow {
  IndexDomain: (L);
  Range: nonnegative;
}
```

```
Variable VoltAngle {
   IndexDomain: (n);
   Range: free;
}
Variable MaxPCC {
   IndexDomain: (c);
   Range: nonnegative;
}
Variable TotalCost {
   Range: free;
   Definition:
w1*(sum(c,CInvstCost(c)*MaxPCC(c)))+w2*(Dur*(sum(g,(ECosta(g)+ECostb(g)*EGenOP(g)+ECostc(g)*EGenOP(g)^2))
))+w3*(Dur*(sum(g,(EEa(g)+EEb(g)*EGenOP(g)+EEc(g)*EGenOP(g)^2))))
}
MathematicalProgram NetCost {
   Objective: TotalCost;
   Direction: minimize;
   Constraints: AllConstraints;
   Variables: AllVariables;
   Type: Automatic;
}
Constraint Nodal {
   IndexDomain: (n);
   Definition:                    sum(g|MapEGen(g,n),EGenOP(g))+sum(c|MapCGen(c,n),CGenOP(c))
sum(L|MapSL(L,n),Flow(L))+sum(L|MapRL(L,n),Flow(L))=sum(d|MapDem(d,n),Load(d));
}
Constraint CBdgtLimit {
   Definition: sum(c,CInvstCost(c)*MaxPCC(c))<=CBdgt;
}
Constraint CCLimit {
   IndexDomain: (c);
   Definition: {
      MaxPCC(c)=sum(q,CB(c,q)*CGenCap(c,q));
   }
}
Constraint CC2 {
   IndexDomain: (c);
   Definition: sum(q,CB(c,q))=1;
}
Constraint EGenLimit {
   IndexDomain: (g);
   Definition: {
      0<=EGenOP(g)<=EGenCap(g);
   }
}
Constraint MaxPCCLimit {
   IndexDomain: (c);
   Definition: {
      0<=MaxPCC(c)<=UPBCC(c);
   }
}
Constraint CGenLimit {
   IndexDomain: (c);
   Definition: {
      CGenOP(c)<=MaxPCC(c);
   }
}
Constraint AngleLimit {
```

```
    IndexDomain: (n);
    Definition: -3.14<=VoltAngle(n)<=3.14;
}
Constraint LineLimit1 {
    IndexDomain: (L);
    Definition: -LineCap(L)<=Flow(L);
}
Constraint LineLimit2 {
    IndexDomain: (L);
    Definition: Flow(L)<=LineCap(L);
}
Constraint FlowLimit1 {
    IndexDomain: (L);
    Definition: {
        Flow(L)=LineSusc(L)*(sum(n|MapSL(L,n),VoltAngle(n))-sum(n|MapRL(L,n),VoltAngle(n)));
    }
}
Procedure MainInitialization {
    Comment: "Add initialization statements here that do NOT require any library being initialized already.";
}
Procedure PostMainInitialization {
    Comment: {
        "Add initialization statements here that require that the libraries are already initialized properly,
        or add statements that require the Data Management module to be initialized."
    }
}
Procedure MainExecution {
    Body: {
        solve NetCost;
    }
 }
```

The simulated results as obtained from the solver are presented in Tables (10.3, 10.4, 10.5 and 10.6). Table 10.3 provides results on the optimal fossil fuel-generated power output, which is lower compared to results reflected in Table 10.3. Table 10.4 gives the optimal sizing and location for renewable energy-generating unit power output. Table 10.5 depicts the OPF on the transmission line and the voltage angle obtainable on each bus is presented in Table 10.6.

**Table 10.3** Optimal generated power from fossil fuel generators for static DC GCEP model

| Generators | Buses | Optimal power (MW) |
|---|---|---|
| G1 | 1 | 139.69 |
| G2 | 3 | 285.36 |
| G3 | 6 | 339.94 |

**Table 10.4** Optimal generated power from prospective RES for static DC GCEP model

| Renewable energy source | Buses | Optimal power (MW) |
|---|---|---|
| RES 1 | 1 | 220 |
| RES 2 | 2 | 280 |
| RES 3 | 4 | 75 |

**Table 10.5** Optimal power flow on the transmission lines for static DC GCEP model

| Between buses | Optimal flow |
|---|---|
| 1–2 | 20.85 |
| 1–4 | 28.84 |
| 1–5 | 150.00 |
| 2–3 | 95.85 |
| 2–4 | 66.71 |
| 2–6 | 150 |
| 3–5 | 150 |
| 3–6 | 40.49 |
| 4–6 | 149.45 |

**Table 10.6** Voltage phase angle at the buses for static DC GCEP model

| Buses | Phase angle (Rad) |
|---|---|
| 1 | 2.76 |
| 2 | 2.68 |
| 3 | 3.06 |
| 4 | 2.54 |
| 5 | 2.46 |
| 6 | 3.14 |

## *10.3.2 Nomenclature*

### Set

g existing generator
k demand
l transmission line
$\Re$ prospective RES

### Parameters

| | |
|---|---|
| $C^{\mathrm{i}}_{\Re}$ | investment cost of the prospective RES $\Re$ |
| $C^{\mathrm{o}}_{\mathrm{g}}$ | operational cost of the existing generator g |
| $E_{\mathrm{g}}$ | emissions cost of the existing generator g |
| $\beta$ | duration factor |
| $P_{\mathrm{k}}$ | system load demand k |
| $P^{\min}_{\mathrm{g}}$ and $P^{\max}_{\mathrm{g}}$ | minimum and maximum power capacity of the existing generator g |
| $\overline{P_{\Re}}$ | upper bound of prospective RES $\Re$ |
| $UB_{\Re}$ | utility budget for prospective RES $\Re$ |
| $P_{\Re c}$ | generation capacity of RES $\Re$ at investment choice c |
| $B_{\mathrm{l}}$ | susceptance of transmission line |

**Variables**

| | |
|---|---|
| $P_{\Re}^{\max}$ | capacity of prospective renewable energy source $\Re$ |
| $P_{g}$ | power produced by the existing generator g |
| $P_{\Re}$ | power produced form the RES $\Re$ |
| $P_{l}$ | power flowing on transmission line l |
| $P_{l}^{+}$ and $P_{l}^{-}$ | power flowing into and out of the transmission line l |
| $\alpha_{\Re c}$ | binary variable for choosing prospective RES $\Re$ at option c |
| $\phi_{b}$ | voltage angle at bus b |

## 10.4 Dynamic Model

Here, a GCEP problem considering a multi-stage approach is examined. This approach involves planning a horizon over a long period; that is, making expansion planning decisions at different times [9]. In this section, the objective function comes with an integrated yearly representation of investment decisions. The objective function seeks to minimize the investment cost of prospective renewable energy-generating units, operational cost, and emissions for the existing fossil fuel-generating-units in the system. The dynamic approach for the GCEP optimization problem can be formulated as:

$$Min\left[\frac{1}{(1+r)^{t}}\left(\sum_{t=1}^{n_t}\sum_{\Re=1}^{n_\Re} C_{\Re}^{i}P_{\Re,t}^{\max}+\beta\sum_{t=1}^{n_t}\sum_{g=1}^{n_g} C_{g}^{o}P_{g,t}+\beta\sum_{t=1}^{n_t}\sum_{g=1}^{n_g} E_{g}P_{g,t}\right)\right] \tag{10.14}$$

$$C_{g}^{o}P_{g,t}=\alpha_{g}+\beta_{g}P_{g,t}+\delta_{g}P_{g,t}^{2} \tag{10.15}$$

$$E_{g}P_{g,t}=a_{g}+b_{g}P_{g,t}+c_{g}P_{g,t}^{2} \tag{10.16}$$

subject to the following constraints:

$$\sum_{g}P_{g,t}+\sum_{\Re}P_{\Re,t}-\sum_{l\in sl}P_{l,t}^{+}+\sum_{r\in rl}P_{l,t}^{-}=\sum_{k}P_{k,t}\forall b,t \tag{10.17}$$

$$\sum_{\Re}C_{\Re}^{i}P_{\Re,t}^{\max}\leq UB_{\Re,t}\forall t \tag{10.18}$$

$$0\leq P_{\Re,t}^{\max}\leq\overline{P_{\Re}}\forall\Re,t \tag{10.19}$$

$$P_{\Re,t}^{\max}=\sum_{c}\alpha_{\Re ct}P_{\Re ct}\forall\Re,t \tag{10.20}$$

$$\sum_{c} \alpha_{\Re ct} = 1 \forall \Re, t \tag{10.21}$$

$$P_{l,t} = B_l \left( \phi_{l \in sl,t} - \phi_{l \in rl,t} \right) \forall l, t \tag{10.22}$$

$$-P_l^{max} \leq P_{l,t} \leq P_l^{max} \forall l, t \tag{10.23}$$

$$0 \leq P_{\Re,t} \leq P_{\Re}^{max} \forall \Re, t \tag{10.24}$$

$$P_g^{min} \leq P_{g,t} \leq P_g^{max} \forall g, t \tag{10.25}$$

$$-\pi \leq \phi_{b,t} \leq \pi \forall b, t. \tag{10.26}$$

The desired variable results to be obtained by the optimization model are $P_{\Re,t}$, $P_{\Re,t}^{max}$, $\phi_{b,t}$, $P_{g,t}$, and $P_{l,t}$.

The objective functions are similar to what is obtained in Sect. 10.3.1, only for the integrated yearly representation of investment decisions in the objective function. Therefore, Eq. (10.14) is categorized into two, namely:

(a) $\sum_{t=1}^{n_t} \sum_{\Re=1}^{n_\Re} C_{\Re}^{i} P_{\Re,t}^{max}$ represents the yearly representation of the annual investment cost of constructing new renewable energy-generating units.
(b) $\beta \sum_{t=1}^{n_t} \sum_{g=1}^{n_g} C_g^{o} P_{g,t}$ and $\beta \sum_{t=1}^{n_t} \sum_{g=1}^{n_g} E_g P_{g,t}$ represent the yearly representation of the cost and emission of operating the generating units.

The investment cost for the prospective renewable energy-generating units is annualized; therefore, a duration factor (8760 h) is multiplied with operational cost and emission of the fossil fuel-generating units to make it comparable with the first objective function. Equations (10.15) and (10.16) are similar to Eqs. (10.2) and (10.3) as the quadratic cost functions for fuel cost and emissions.

Constraint (10.17) states that power generated by all the generating units (fossil fuel and renewable energy generators) minus the power flowing in and out of the transmission lines should be equal to the power demanded by the load. Constraint (10.18) is the renewable energy investment budget and Eqs. (10.19) to (10.21) are the constraints for sizing the renewable energy-generating units. Equation (10.22) is the constraint for flow limit on the transmission line and this is affected by the susceptance and voltage angle of the transmission line and constraint (10.23) is the limit for the transmission line. Equations (10.24) and (10.25) are the generation limits for the fossil fuel and renewable energy generators. Constraint (10.26) is the voltage angle bound limit [10].

The performance of the developed dynamic GCEP model is validated on the modified IEEE six-bus test system as presented in Fig. 10.1. The data for the test bus system can be obtained in Tables 10.1 and 10.2, except for the yearly load profile at the load bus, which is depicted in Table 10.7. The planning period horizon is for a period of 10 years.

The AIMMS code for solving the dynamic DC GCEP model is given below:

```
Model Main_newwork {
  Set generator {
    Index: g;
  }
  Set nodes {
    Index: n;
  }
  Set demand {
    Index: d;
  }
  Set periods {
    Index: t;
  }
  Set line {
    Index: l;
  }
  Set el {
    SubsetOf: line;
  }
  Set cl {
    SubsetOf: line;
  }
  Set Res {
    Index: R;
  }
  Set Ropt {
    Index: q;
  }
  Parameter mapgen {
    IndexDomain: (g,n);
  }
  Parameter mapRes {
    IndexDomain: (R,n);
  }
  Parameter mapdem {
    IndexDomain: (d,n);
  }
  Parameter mapsl {
    IndexDomain: (l,n);
  }
  Parameter maprl {
    IndexDomain: (l,n);
  }
  Parameter demands {
    IndexDomain: (d,t);
  }
  Parameter fmax {
    IndexDomain: l;
  }
  Parameter bb {
    IndexDomain: l;
  }
  Parameter pmax {
    IndexDomain: g;
  }
  Parameter aa {
    IndexDomain: g;
  }
```

```
Parameter b {
  IndexDomain: g;
}
Parameter c {
  IndexDomain: g;
}
Parameter ea {
  IndexDomain: g;
}
Parameter eb {
  IndexDomain: g;
}
Parameter ec {
  IndexDomain: g;
}
Parameter investl {
  IndexDomain: l;
}
Parameter investR {
  IndexDomain: R;
}
Parameter M;
Parameter dur;
Parameter k;
Parameter kk {
  IndexDomain: t;
  Definition: (1/(1+0.085)^v(t));
}
Parameter v {
  IndexDomain: t;
}
Parameter ResCap {
  IndexDomain: (t,R,q);
}
Parameter URES {
  IndexDomain: (R);
}
Parameter u1;
Parameter u2;
Parameter u3;
Parameter u4;
Parameter u5;
Variable z {
  IndexDomain: (l,t);
  Range: binary;
}
Variable Rz {
  IndexDomain: (t,R,q);
  Range: binary;
}
Variable RMax {
  IndexDomain: (R,t);
  Range: nonnegative;
}
Variable Rp {
  IndexDomain: (R,t);
  Range: nonnegative;
}
```

```
Variable pg {
  IndexDomain: (g,t);
  Range: nonnegative;
}
Variable flow {
  IndexDomain: (l,t);
  Range: nonnegative;
}
Variable ang {
  IndexDomain: (n,t);
  Range: free;
}
Variable Totalcost {
  Range: free;
  Definition: sum(t,(1/(1+k)^v(t))*((u1*sum(R, investR(R)*RMax(R,t)))+(u2*dur*sum(g,
(aa(g)+(b(g)*pg(g,t))+(c(g)*pg(g,t)^2))))+(u3*dur*sum(g, (ea(g)+(eb(g)*pg(g,t))+(ec(g)*pg(g,t)^2))))-(u4*0.3*(sum(R,
investR(R)*RMax(R,t))))));
}
MathematicalProgram Leastcost {
  Objective: Totalcost;
  Direction: minimize;
  Constraints: AllConstraints;
  Variables: AllVariables;
  Type: Automatic;
}
Constraint powerbalance {
  IndexDomain: (n,t);
  Definition: sum(g|mapgen(g,n),pg(g,t))+sum(R|mapRes(R,n),Rp(R,t))-
sum(l|mapsl(l,n),flow(l,t))+sum(l|maprl(l,n),flow(l,t))=sum(d|mapdem(d,n),demands(d,t));
}
Constraint linelimit1 {
  IndexDomain: (l,t);
  Definition: -fmax(l)<=flow(l,t);
}
Constraint linelimit2 {
  IndexDomain: (l,t);
  Definition: flow(l,t)<=fmax(l);
}
Constraint genlimit1 {
  IndexDomain: (g,t);
  Definition: 0<=pg(g,t);
}
Constraint genlimit2 {
  IndexDomain: (g,t);
  Definition: pg(g,t)<=pmax(g);
}
Constraint buslimit {
  IndexDomain: (n,t);
  Definition: -3.14<=ang(n,t)<=3.14;
}
Constraint flowlimit1 {
  IndexDomain: (l,t);
  Definition: flow(l,t)=bb(l)*(sum(n|mapsl(l,n),ang(n,t))-sum(n|maprl(l,n),ang(n,t)));
}
Constraint Rlimit {
  IndexDomain: (R,t);
  Definition: RMax(R,t)=sum(q,Rz(t,R,q)*ResCap(t,R,q));
}
```

```
Constraint Rlimit2 {
   IndexDomain: (R,t);
   Definition: sum(q,Rz(t,R,q))=1;
}
Constraint Rlimit3 {
   IndexDomain: (R,t);
   Definition: Rp(R,t)<=RMax(R,t);
}
Constraint Rlimit4 {
   IndexDomain: (R,t);
   Definition: 0<=RMax(R,t)<=URES(R);
}
Constraint nn {
   IndexDomain: (R,t)|ord(t)<=9;
   Definition: Rp(R,t+1)>=Rp(R,t);
}
Procedure MainInitialization {
   Comment: "Add initialization statements here that do NOT require any library being initialized already.";
}
Procedure PostMainInitialization {
   Comment: {
      "Add initialization statements here that require that the libraries are already initialized properly,
      or add statements that require the Data Management module to be initialized."
   }
}
Procedure MainExecution {
   Body: {
      solve Leastcost;
   }
}
```

**Table 10.7** Modified load profile for dynamic DC GCEP model

| Buses | Period (year) | | | | | | | | | |
|---|---|---|---|---|---|---|---|---|---|---|
| | 1 | 2 | 3 | 4 | 5 | 6 | 7 | 8 | 9 | 10 |
| 1 | 80 | 88 | 96 | 104 | 112 | 120 | 128 | 136 | 144 | 160 |
| 2 | 240 | 264 | 288 | 312 | 336 | 360 | 384 | 408 | 432 | 480 |
| 3 | 40 | 44 | 48 | 52 | 56 | 60 | 64 | 68 | 72 | 80 |
| 4 | 160 | 176 | 192 | 208 | 224 | 240 | 256 | 272 | 288 | 320 |
| 5 | 240 | 264 | 288 | 300 | 300 | 300 | 300 | 300 | 300 | 300 |

The simulated results as obtained from the solver are presented in Tables (10.8, 10.9, 10.10, 10.11). Tables 10.8 and 10.9 show the optimal power produced by fossil fuel and renewable energy generators at different periods of the year. It was observed that the capacity of the fossil fuel generators reduced owing to incorporation of more renewable energy into the system as the period of planning increased. The yearly representation of the OPF on the transmission line is presented in Tables 10.10 and 10.11 gives the bus voltage angles at each bus.

**Table 10.8** Optimal power from fossil fuel generators for dynamic DC GCEP model

| Generators | Period (year) | | | | | | | | | |
|---|---|---|---|---|---|---|---|---|---|---|
| | 1 | 2 | 3 | 4 | 5 | 6 | 7 | 8 | 9 | 10 |
| 1 | 150.00 | 147.34 | 132.43 | 142.84 | 125.73 | 135.65 | 146.07 | 125.44 | 135.85 | 127.16 |
| 2 | 18.26 | 47.00 | 142.80 | 194.80 | 199.81 | 202.80 | 206.93 | 210.80 | 214.80 | 222.80 |
| 3 | 411.74 | 428.84 | 393.79 | 374.32 | 372.46 | 377.71 | 381.99 | 375.45 | 379.98 | 380.55 |

**Table 10.9** Optimal power from renewable energy generators for dynamic DC GCEP model

| Renewable energy sources | Period (year) | | | | | | | | | |
|---|---|---|---|---|---|---|---|---|---|---|
| | 1 | 2 | 3 | 4 | 5 | 6 | 7 | 8 | 9 | 10 |
| 1 | 100 | 100 | 125 | 125 | 150 | 150 | 150 | 175 | 175 | 200 |
| 2 | 80.00 | 112.82 | 117.98 | 139.04 | 160 | 193.83 | 227.01 | 237.30 | 270.36 | 319.50 |
| 3 | – | – | – | – | 20 | 20 | 20 | 60 | 60 | 90 |

**Table 10.10** Optimal power flow on the transmission lines for dynamic DC GCEP model

| Lines | Period (year) | | | | | | | | | |
|---|---|---|---|---|---|---|---|---|---|---|
| | 1 | 2 | 3 | 4 | 5 | 6 | 7 | 8 | 9 | 10 |
| 1–2 | – | – | – | – | – | – | – | – | – | 11.64 |
| 1–4 | 13 | 9.34 | 11.43 | 13.84 | 13.39 | 15.65 | 18.07 | 14.44 | 16.85 | 17.16 |
| 1–5 | 145.35 | 150 | 150 | 150 | 150 | 150 | 150 | 150 | 150 | 150 |
| 2–3 | 25.59 | 35.11 | 42.98 | 52.04 | 49.67 | 58.83 | 67.93 | 54.30 | 63.36 | 64.50 |
| 2–4 | 33.61 | 39 | 63 | 75 | 75.34 | 75 | 75 | 75 | 75 | 75 |
| 2–6 | 94.65 | 114 | 138 | 150 | 150 | 150 | 150 | 150 | 150 | 150 |
| 3–5 | 140.33 | 147.29 | 150 | 150 | 150 | 150 | 149.92 | 150 | 150 | 150 |
| 3–6 | 150 | 150 | 106.20 | 82.20 | 81.53 | 82.20 | 82.07 | 82.20 | 82.20 | 82.20 |
| 4–6 | 121.41 | 131.55 | 137.59 | 142.12 | 140.94 | 145.52 | 150.00 | 143.25 | 147.78 | 148.35 |

**Table 10.11** Voltage phase angle for dynamic DC GCEP model

| Buses | Phase angle (Rad) |
|---|---|
| 1 | 2.68 |
| 2 | 2.68 |
| 3 | 2.93 |
| 4 | 2.59 |
| 5 | 2.38 |
| 6 | 3.14 |

## *10.4.1 Nomenclature*

### Set

g existing generator
k demand
$t$ time
$\Re$ prospective RES

**Parameters**

| | |
|---|---|
| $C^{\mathrm{i}}_{\Re}$ | investment cost of the prospective RES $\Re$ |
| $C^{\mathrm{o}}_{\mathrm{g}}$ | operation cost of the existing generator g |
| $E_{\mathrm{g}}$ | emissions cost of the existing generator g |
| $\beta$ | duration factor |
| $P_{\mathrm{k},t}$ | system load demand k at time |
| $P^{\min}_{\mathrm{g}}$ and $P^{\max}_{\mathrm{g}}$ | minimum and maximum power capacity of the existing generator g |
| $\overline{P_{\Re}}$ | upper bound of the prospective RES $\Re$ |
| $UB_{\Re,t}$ | utility budget for prospective RES $\Re$ at time |
| $P_{\Re ct}$ | generation capacity of RES $\Re$ at investment choice c |
| $B_{\mathrm{l}}$ | susceptance of transmission line |

**Variables**

| | |
|---|---|
| $P^{\max}_{\Re,t}$ | capacity of prospective renewable energy source $\Re$ at time |
| $P_{\mathrm{g},t}$ | power produced by the existing generator g at time |
| $P_{\Re,t}$ | power produced by the RES $\Re$ at time |
| $P_{\mathrm{l},t}$ | power flowing on transmission line l at time t |
| $P^{+}_{\mathrm{l},t}$ and $P^{-}_{\mathrm{l},t}$ | power flowing into and out of the transmission line l at time $t$ |
| $\alpha_{\Re ct}$ | binary variable for choosing prospective RES $\Re$ at option c |
| $\phi_{\mathrm{b},t}$ | voltage angle at bus b at time t |

## 10.5 Reactive Power Capacity Expansion Planning

Here, we consider the reactive power of the GCEP problem. In Sect. 10.5.3, the active power of GCEP ignores bus voltage and reactive power. On the contrary, the reactive power of GCEP considers the bus voltage and reactive power using an AC model to assess the real and reactive power losses for basic configuration and proper planning systems accurately [11, 12]. This section provides solutions for reactive power of GCEP problems for an AC model in AIMMS. Table 10.12 presents the real and reactive parameters of the generator and load.

Here, the objective function seeks to minimize the investment cost of prospective renewable energy-generating units, operational cost, and emissions for the existing fossil fuel-generating units in the system for a GCEP static model while satisfying the AC constraints.

$$Min\left[\sum_{\Re=1}^{n_{\Re}} C^{\mathrm{i}}_{\Re} P^{\max}_{\Re} + \beta \sum_{g=1}^{n_g} C^{\mathrm{o}}_{\mathrm{g}} P_{\mathrm{g}} + \beta \sum_{g=1}^{n_g} E_{\mathrm{g}} P_{\mathrm{g}}\right] \tag{10.27}$$

$$C^{\mathrm{o}}_{\mathrm{g}} P_{\mathrm{g}} = \alpha_{\mathrm{g}} + \beta_{\mathrm{g}} P_{\mathrm{g}} + \delta_{\mathrm{g}} P^{2}_{\mathrm{g}} \tag{10.28}$$

**Table 10.12** Real and reactive parameters for IEEE Garver six-bus system example

| Bus | Parameters of generators | | | | Load parameter | |
|---|---|---|---|---|---|---|
| | $P_g^{min}$(MW) | $P_g^{max}$ (MW) | $Q_g^{min}$ (MVAr) | $Q_g^{max}$ (MVAr) | $P_k$ (MW) | $Q_k$ (MVAr) |
| 1 | 0 | 160 | −10 | 65 | 80 | 16 |
| 2 | – | – | – | – | 240 | 48 |
| 3 | 0 | 360 | −10 | 150 | 40 | 8 |
| 4 | – | – | – | – | 160 | 32 |
| 5 | – | – | – | – | 240 | 48 |
| 6 | 0 | 600 | −10 | 200 | – | – |

$$E_g P_g = a_g + b_g P_g + c_g P_g^2 \tag{10.29}$$

subject to the following constraints:

$$\sum_{g} P_g + \sum_{\Re} P_{\Re} - \sum_{l \in sl} P_{loss} = \sum_{k} P_k \forall b \tag{10.30}$$

$$\sum_{g} Q_g + \sum_{\Re} Q_{\Re} - \sum_{l \in sl} Q_{loss} = \sum_{k} Q_k \forall b \tag{10.31}$$

$$\sum_{\Re} C_{\Re}^{i} P_{\Re}^{max} \le UB_{\Re} \tag{10.32}$$

$$0 \le P_{\Re}^{max} \le \overline{P_{\Re}} \forall \Re \tag{10.33}$$

$$P_{\Re}^{max} = \sum_{c} \alpha_{\Re c} P_{\Re c} \forall \Re \tag{10.34}$$

$$\sum_{c} \alpha_{\Re c} = 1 \forall \Re \tag{10.35}$$

$$P_l^{+} = V_a g_l - V_a V_b (g_l \cos\theta_{ab} + b_l \sin\theta_{ab}) \forall l \tag{10.36}$$

$$Q_l^{+} = -V_a (b_l + b_{lo}) + V_a V_b (g_l \cos\theta_{ab} - \mathrm{b}_l \sin\theta_{ab}) \forall l \tag{10.37}$$

$$P_l^{-} = V_b g_l - V_a V_b (g_l \cos\theta_{ab} - \mathrm{b}_l \sin\theta_{ab}) \forall l \tag{10.38}$$

$$Q_l^{-} = -V_b (b_l + b_{lo}) + V_a V_b (b_l \cos\theta_{ab} + \mathrm{g}_l \sin\theta_{ab}) \forall l \tag{10.39}$$

$$P_{\text{loss}} = g_l(V_a^2 + V_b^2 - 2V_aV_b\cos\theta_{ab})\forall l \tag{10.40}$$

$$Q_{\text{loss}} = -b_{lo}(V_a^2 + V_b^2) - \mathrm{b}_l(V_a^2 + V_b^2 - 2V_aV_b\cos\theta_{ab})\forall l \tag{10.41}$$

$$0 \le P^{\Re} \le P_{\Re}^{\max}\forall\Re \tag{10.42}$$

$$Q_{\Re}^{\min} \le Q^{\Re} \le Q_{\Re}^{\max}\forall\Re \tag{10.43}$$

$$P_g^{\min} \le P_g \le P_g^{\max}\forall g \tag{10.44}$$

$$Q_g^{\min} \le Q_g \le Q_g^{\max}\forall g \tag{10.45}$$

$$0 \le (P_l^+)^2 + (Q_l^+)^2 \le S_l^2\forall l \tag{10.46}$$

$$0 \le (P_l^-)^2 + (Q_l^-)^2 \le S_l^2\forall l \tag{10.47}$$

$$V_{\min} \le V_b \le V_{\max}\forall b \tag{10.48}$$

$$-\pi \le \theta_b \le \pi\forall b. \tag{10.49}$$

The desired variable results to be obtained by the optimization model are $P_{\Re}$, $P_{\Re}^{\max}$, $\phi_{\mathrm{b}}$, $V_{\mathrm{b}}$, $P_{\mathrm{g}}$, $Q_{\mathrm{g}}$, $P_1^+$, and $P_1^-$.

The objective functions are similar to what is obtained in Sect. 10.3.1, the only difference being that the model here is an AC model. Therefore, Eqs. (10.27–10.29) are similar to Eqs. (10.14–10.16).

Equations (10.30) and (10.31) are the real and reactive power constraints that ensure that the real and reactive power produced by the fossil fuel generators, renewable energy generators and the real and reactive power flowing in the lines equal the demand of the consumers. Equations (10.32) to (10.35) define the optimal rating capacity of the new renewable energy generators. Constraints (10.36) to (10.39) describe the real and reactive power flowing into and out of the lines, which is a function of the bus voltage and line parameters. Equations (10.40) and (10.41) define the real and reactive power loss constraints in the line, which ensure that the real and reactive power loss are equal to the total power flowing into and out of the lines. Constraints (10.42) to (10.45) are the real and reactive limits for the renewable energy and fossil fuel generators, which must not exceed their bounds. Equations (10.46) and (10.47) are the capacity limit for the lines. Constraints (10.48) and (10.49) are the voltage and angle limits for the bus.

The AIMMS code for solving AC GCEP model is given below:

```
Model Main_ACLoss {
  Set Nodes {
    Index: n;
  }
  Set Lines {
    Index: L;
  }
  Set EGen {
    Index: g;
  }
  Set CGen {
    Index: c;
  }
  Set SGen {
    Index: s;
  }
  Set Demand {
    Index: d;
  }
  Set ELines {
    SubsetOf: Lines;
  }
  Set CLines {
    SubsetOf: Lines;
  }
  Set RefNode {
    SubsetOf: Nodes;
  }
  Set ProdOption {
    Index: q;
  }
  Set harm {
    Index: h;
  }
  Parameter QCGenCap {
    IndexDomain: (c,q);
  }
  Parameter UQCGenn {
    IndexDomain: (c);
  }
  Parameter EGenMap {
    IndexDomain: (g,n);
  }
  Parameter CGenMap {
    IndexDomain: (c,n);
  }
  Parameter DemMap {
    IndexDomain: (d,n);
  }
  Parameter MapSL {
    IndexDomain: (L,n);
  }
  Parameter MapRL {
    IndexDomain: (L,n);
  }
  Parameter RLoad {
    IndexDomain: (d);
```

```
}
Parameter QLoad {
  IndexDomain: (d);
}
Parameter REGenCap {
  IndexDomain: (g);
}
Parameter QEGenCapMx {
  IndexDomain: (g);
}
Parameter QEGenCapMn {
  IndexDomain: (g);
}
Parameter SCap {
  IndexDomain: (L);
}
Parameter EGenCost {
  IndexDomain: (g);
}
Parameter CGenCost {
  IndexDomain: (c);
}
Parameter InvestCost {
  IndexDomain: (c);
}
Parameter InvestLCost {
  IndexDomain: (L);
}
Parameter WegFac;
Parameter UpperBoundCCap {
  IndexDomain: (c);
}
Parameter LineSusc {
  IndexDomain: (L);
}
Parameter LineSusc2 {
  IndexDomain: (L);
}
Parameter M;
Parameter Conduct {
  IndexDomain: (L);
}
Parameter Vmin {
  IndexDomain: (n);
}
Parameter Vmax {
  IndexDomain: (n);
}
Parameter CGenCap {
  IndexDomain: (c,q);
}
Parameter ADD {
  Definition: {
     sum(L,Conduct(L)*(sum(n|MapSL(L,n),sqr(Voltage(n)))+sum(n|MapRL(L,n),sqr(Voltage(n)))-
2*(sum(n|MapSL(L,n),Voltage(n))*sum(n|MapRL(L,n),Voltage(n))*cos(sum(n|MapSL(L,n),Angle(n))-
sum(n|MapRL(L,n),Angle(n))))));
  }
```

```
}
Parameter w1;
Parameter w2;
Parameter w3;
Parameter w4;
Parameter Vh {
  IndexDomain: (n,h);
}
Parameter RL {
  IndexDomain: (L);
}
Parameter XL {
  IndexDomain: (L);
}
Parameter RE {
  IndexDomain: (L,h);
}
Parameter XE {
  IndexDomain: (L,h);
}
Parameter YE {
  IndexDomain: (L,h);
}
Parameter YN {
  IndexDomain: (n,h);
}
Parameter a {
  IndexDomain: (n);
}
Parameter b {
  IndexDomain: (n);
}
Parameter cc {
  IndexDomain: (n);
}
Parameter costa {
  IndexDomain: g;
}
Parameter costb {
  IndexDomain: g;
}
Parameter costc {
  IndexDomain: g;
}
Parameter Ea {
  IndexDomain: g;
}
Parameter Eb {
  IndexDomain: g;
}
Parameter Ec {
  IndexDomain: g;
}
Variable Pn {
  IndexDomain: (n);
  Range: free;
}
```

```
Variable RFL_SN {
  IndexDomain: (L);
  Range: nonnegative;
}
Variable QFL_SN {
  IndexDomain: (L);
  Range: free;
}
Variable RFL_RN {
  IndexDomain: (L);
  Range: free;
}
Variable QFL_RN {
  IndexDomain: (L);
  Range: free;
}
Variable BD {
  IndexDomain: (L);
  Range: binary;
}
Variable UOption {
  IndexDomain: (c,q);
  Range: binary;
}
Variable CGenOP {
  IndexDomain: (c);
  Range: nonnegative;
}
Variable QCGenOP {
  IndexDomain: (c);
  Range: free;
}
Variable MaxPCGen {
  IndexDomain: (c);
  Range: nonnegative;
}
Variable Voltage {
  IndexDomain: (n);
  Range: nonnegative;
}
Variable REGenOP {
  IndexDomain: (g);
  Range: nonnegative;
}
Variable QEGenOP {
  IndexDomain: (g);
  Range: free;
}
Variable RPLoss {
  IndexDomain: (L);
  Range: free;
}
Variable QPLoss {
  IndexDomain: (L);
  Range: free;
}
Variable AB {
  IndexDomain: (L);
```

```
    Range: nonnegative;
    Definition: QFL_SN(L)+QFL_RN(L);
  }
  Variable Angle {
    IndexDomain: (n);
    Range: free;
  }
  Variable Total {
    Range: free;
    Definition:
w1*(sum(c,InvestCost(c)*MaxPCGen(c)))+w2*(WegFac*(sum(g,(costa(g)+costb(g)*REGenOP(g)+costc(g)*REGenOP(g)^
2))))+w3*(WegFac*(sum(g,(Ea(g)+Eb(g)*REGenOP(g)+Ec(g)*REGenOP(g)^2))));
  }
  MathematicalProgram NetCost {
    Objective: Total;
    Direction: minimize;
    Constraints: AllConstraints;
    Variables: AllVariables;
    Type: Automatic;
  }
  ElementParameter myGMP {
    Range: AllGeneratedMathematicalPrograms;
  }
  Constraint C1 {
    IndexDomain: (c);
    Definition: {
      MaxPCGen(c)=sum(q,UOption(c,q)*CGenCap(c,q));
    }
  }
  Constraint C2 {
    IndexDomain: (c);
    Definition: sum(q,UOption(c,q))=1;
  }
  Constraint C3 {
    IndexDomain: (c);
    Definition: {
      0<=MaxPCGen(c)<=UpperBoundCCap(c);
    }
  }
  Constraint C4 {
    IndexDomain: (c);
    Definition: {
      CGenOP(c)<=MaxPCGen(c);
    }
  }
  Constraint VL1 {
    IndexDomain: (n);
    Definition: {
      Vmin(n)<=Voltage(n);
    }
  }
  Constraint VL2 {
    IndexDomain: (n);
    Definition: {
      Voltage(n)<=Vmax(n);
    }
  }
```

```
Constraint A1 {
   IndexDomain: (n)|ord(n)=1;
   Definition: Angle(n)=0;
}
Constraint A2 {
   IndexDomain: (n);
   Definition: {
      -3.14<=Angle(n);
   }
}
Constraint A3 {
   IndexDomain: (n);
   Definition: Angle(n)<=3.14;
}
Constraint n1 {
   IndexDomain: (n);
   Definition: {
      sum(g|EGenMap(g,n),REGenOP(g))+sum(c|CGenMap(c,n),CGenOP(c))-sum(L|MapSL(L,n),RFL_SN(L))-
sum(L|MapRL(L,n),RFL_RN(L))=sum(d|DemMap(d,n),RLoad(d));
   }
}
Constraint n2 {
   IndexDomain: (n);
   Definition:                    sum(g|EGenMap(g,n),QEGenOP(g))+sum(c|CGenMap(c,n),QCGenOP(c))-
sum(L|MapSL(L,n),QFL_SN(L))-sum(L|MapRL(L,n),QFL_RN(L))=sum(d|DemMap(d,n),QLoad(d));
}
Constraint FS1 {
   IndexDomain: (L);
   Definition:                    RFL_SN(L)=(sqr(sum(n|MapSL(L,n),Voltage(n)))*Conduct(L))-
(sum(n|MapSL(L,n),Voltage(n))*sum(n|MapRL(L,n),Voltage(n)))*(Conduct(L)*(cos(sum(n|MapSL(L,n),Angle(n))-
sum(n|MapRL(L,n),Angle(n))))+(-LineSusc(L)*sin(sum(n|MapSL(L,n),Angle(n))-sum(n|MapRL(L,n),Angle(n)))));
}
Constraint FS11 {
   IndexDomain: (L);
   Definition:                    QFL_SN(L)=-(sqr(sum(n|MapSL(L,n),Voltage(n))))*(LineSusc2(L)+(-
LineSusc(L)))+((sum(n|MapSL(L,n),Voltage(n))*sum(n|MapRL(L,n),Voltage(n)))*(((-
LineSusc(L))*(cos(sum(n|MapSL(L,n),Angle(n))-sum(n|MapRL(L,n),Angle(n)))))-
(Conduct(L)*(sin(sum(n|MapSL(L,n),Angle(n))-sum(n|MapRL(L,n),Angle(n))))));
}
Constraint FR1 {
   IndexDomain: (L);
   Definition:                    RFL_RN(L)=(sqr(sum(n|MapRL(L,n),Voltage(n)))*Conduct(L)-
((sum(n|MapSL(L,n),Voltage(n))*sum(n|MapRL(L,n),Voltage(n)))*(Conduct(L)*(cos(sum(n|MapSL(L,n),Angle(n))-
sum(n|MapRL(L,n),Angle(n)))))-(-LineSusc(L)*sin(sum(n|MapSL(L,n),Angle(n))-sum(n|MapRL(L,n),Angle(n))))));
}
Constraint FR11 {
   IndexDomain: (L);
   Definition:                    QFL_RN(L)=(-(sqr(sum(n|MapRL(L,n),Voltage(n))))*(LineSusc2(L)+(-
LineSusc(L)))+((sum(n|MapSL(L,n),Voltage(n))*sum(n|MapRL(L,n),Voltage(n)))*(Conduct(L)*(sin(sum(n|MapSL(L,n),A
ngle(n))-sum(n|MapRL(L,n),Angle(n))))+((-LineSusc(L))*(cos(sum(n|MapSL(L,n),Angle(n))-
sum(n|MapRL(L,n),Angle(n))))))));
}
Constraint G1 {
   IndexDomain: (g);
   Definition: {
      0<=REGenOP(g);
   }
```

```
    }
    Constraint G2 {
        IndexDomain: (g);
        Definition: {
            REGenOP(g)<=REGenCap(g);
        }
    }
    Constraint G3 {
        IndexDomain: (g);
        Definition: {
            QEGenCapMn(g)<=QEGenOP(g);
        }
    }
    Constraint G4 {
        IndexDomain: (g);
        Definition: QEGenOP(g)<=QEGenCapMx(g);
    }
    Constraint QG1 {
        IndexDomain: (c);
        Definition: -UQCGenn(c)<=QCGenOP(c);
    }
    Constraint QG2 {
        IndexDomain: (c);
        Definition: QCGenOP(c)<=UQCGenn(c);
    }
    Constraint PL1 {
        IndexDomain: (L);
        Definition:      RPLoss(L)=Conduct(L)*(sum(n|MapSL(L,n),sqr(Voltage(n)))+sum(n|MapRL(L,n),sqr(Voltage(n)))-
2*(sum(n|MapSL(L,n),Voltage(n))*sum(n|MapRL(L,n),Voltage(n))*cos(sum(n|MapSL(L,n),Angle(n))-
sum(n|MapRL(L,n),Angle(n)))));
    }
    Constraint QL1 {
        IndexDomain: (L);
        Definition:  QPLoss(L)=-(-LineSusc(L))*((sum(n|MapSL(L,n),sqr(Voltage(n)))+sum(n|MapRL(L,n),sqr(Voltage(n))))-
2*(sum(n|MapSL(L,n),Voltage(n))*sum(n|MapRL(L,n),Voltage(n)))*(cos(sum(n|MapSL(L,n),Angle(n))-
sum(n|MapRL(L,n),Angle(n)))));
    }
    Constraint SS1 {
        IndexDomain: (L);
        Definition: 0<=(sqr(RFL_SN(L))+sqr(QFL_SN(L)));
    }
    Constraint SS2 {
        IndexDomain: (L);
        Definition: (sqr(RFL_SN(L))+sqr(QFL_SN(L)))<=sqr(SCap(L));
    }
    Constraint SS3 {
        IndexDomain: (L);
        Definition: (sqr(RFL_RN(L))+sqr(QFL_RN(L)))<=sqr(SCap(L));
    }
    Procedure MainInitialization {
        Comment: "Add initialization statements here that do NOT require any library being initialized already.";
    }
    Procedure PostMainInitialization {
        Comment: {
            "Add initialization statements here that require that the libraries are already initialized properly,
            or add statements that require the Data Management module to be initialized."
        }
```

```
}
Procedure MainExecution {
  Body: {
    !solve NetCost;

    myGMP:=GMP::Instance::Generate( NetCost );

    GMPOuterApprox::CreateStatusFile:=1;

    GMPOuterApprox::IsConvex:=1;

    GMPOuterApprox::DoOuterApproximation( myGMP );
  }
}
```

The performance of the developed AC GCEP model is validated on the modified IEEE six-bus test system as presented in Fig. 10.1. The data for the test bus system can be obtained in Table 10.12. The AIMMS code for solving the AC GCEP model is given as ACode 10.4.

Tables 10.13 and 10.14 present the optimal real and reactive power output from fossil fuel- and renewable energy-generating units, respectively. Table 10.15 describes the optimal power flowing in and out of the transmission lines, as well as the line losses. In Table 10.16, the voltage angle and magnitude for the six-bus test system is presented. The angles are in rad while the magnitudes are per unit.

**Table 10.13** Optimal real and reactive power from fossil fuel generators for AC GCEP model

| Generators | Buses | Optimal real power (MW) | Optimal reactive power (MVar) |
|---|---|---|---|
| G1 | 1 | 160 | 20.11 |
| G2 | 3 | 292.21 | 150 |
| G3 | 6 | 309.54 | 47.59 |

**Table 10.14** Optimal real and reactive power from renewable energy generators for AC GCEP model

| Renewable Energy Source | Buses | Optimal real power (MW) | Optimal reactive power (MVar) |
|---|---|---|---|
| RES 1 | 1 | 260 | 95.02 |
| RES 2 | 2 | 140 | 36.47 |
| RES 3 | 4 | 125 | 33.33 |

**Table 10.15** Optimal power flow and line losses on the transmission lines for AC GCEP model

| Between Buses | Outflow power (MW) | Inflow power (MW) | Line losses (MW) |
|---|---|---|---|
| 1–2 | 72.84 | −70.92 | 2.22 |
| 1–4 | 48.46 | −47.18 | 1.48 |
| 1–5 | 138.69 | −135.19 | 4.07 |
| 2–4 | 123.45 | −121.27 | 3.27 |
| 2–6 | 116.74 | −114.81 | 2.96 |
| 3–5 | 140.36 | −147.82 | 4.25 |
| 3–6 | 28.81 | −27.97 | 0.47 |
| 4–6 | 140.36 | −147.82 | 4.25 |

**Table 10.16** Voltage magnitude and phase angle for AC GCEP model

| Buses | Voltage magnitude (PU) | Phase angle (Rad) |
|---|---|---|
| 1 | 1.048 | 2.70 |
| 2 | 0.95 | 2.62 |
| 3 | 1.043 | 3.02 |
| 4 | 0.95 | 2.57 |
| 5 | 0.97 | 2.76 |
| 6 | 1.048 | 3.14 |

## 10.5.1 Nomenclature

### Set

g existing generator
k demand
l transmission line
$\Re$ prospective RES

### Parameters

$C^{\mathrm{i}}_{\Re}$ investment cost of the prospective RES $\Re$
$C^{\mathrm{o}}_{\mathrm{g}}$ operational cost of the existing generator g
$E_{\mathrm{g}}$ emissions cost of the existing generator g
$\beta$ duration factor
$P_{\mathrm{k}}$ and $Q_{\mathrm{k}}$ active and reactive system load demand k
$P^{\mathrm{min}}_{\mathrm{g}}$ and $Q^{\mathrm{min}}_{\mathrm{g}}$ minimum active and reactive power capacity of the existing generator g
$P^{\mathrm{max}}_{\mathrm{g}}$ and $Q^{\mathrm{max}}_{\mathrm{g}}$ maximum active and reactive power capacity of the existing generator g
$\overline{P_{\Re}}$ upper bound of the prospective RES $\Re$
$S_{\mathrm{l}}$ apparent power of the transmission line l

| | |
|---|---|
| $UB_{\Re}$ | utility budget for prospective RES $\Re$ |
| $P_{\Re c}$ | generation capacity of RES $\Re$ at investment choice c |
| $B_l$ | susceptance of transmission line |
| $V_{min}$ and $V_{max}$ | minimum and maximum voltage magnitude |
| $V_a$ and $V_b$ | sending and receiving end bus, respectively |
| $b_l$, $b_{lo}$ and $g_l$ | susceptance, shunt susceptance, and conductance of transmission line, respectively |

**Variables**

| | |
|---|---|
| $P_{\Re}^{max}$ | capacity of prospective renewable energy source $\Re$ |
| $P_g$ and $Q_g$ | active and reactive power produced by the existing generator g |
| $P_{\Re}$ and $Q_{\Re}$ | active and reactive power produced by the RES $\Re$ |
| $P_l$ and $Q_l$ | active and reactive power flowing on transmission line l |
| $P_{loss}$ and $Q_{loss}$ | active and reactive power loss on transmission line l |
| $P_l^+$ and $Q_l^+$ | active and reactive power flowing into of the transmission line l |
| $P_l^-$ and $Q_l^-$ | active and reactive power flowing out of the transmission line l |
| $\alpha_{\Re c}$ | binary variable for choosing prospective RES $\Re$ at option c |
| $\phi_b$ | voltage angle at bus b |
| $V_b$ | voltage magnitude at bus b |

## 10.6 Demand Response Integrated Generator Capacity Expansion Planning

This section explains the system scheduling between the generation and demand sides. As discussed in Sect. 10.4, DR is regarded as a means of reshaping the system load profile to achieve improved system efficiency. DR can be classified as price-based DR and incentive-based DR [5, 13]. With increased penetration of RES into the grid, DR plays a vital role in power systems, such as smoothing the unstable generation profile of RES owing to the randomness and uncertainty associated with them and consequently reducing the operating costs of the system [14]. Here, incentive-based DR is employed and the mathematical formulation of GCEP with DR is shown below:

$$Min\left[\frac{1}{(1+r)^t}\left(\sum_{t=1}^{n_t}\sum_{\Re=1}^{n_{\Re}}C_{\Re}^{i}P_{\Re,t}^{max}+\beta\sum_{t=1}^{n_t}\sum_{g=1}^{n_g}C_g^{o}P_{g,t}+\beta\sum_{t=1}^{n_t}\sum_{g=1}^{n_g}E_gP_{g,t}+\lambda\sum_{t=1}^{n_t}\sum_{y=1}^{n_y}\delta_y(C_y^{r}\alpha_{y,t}^{r})\right)\right] \tag{10.50}$$

$$C_g^{o}P_{g,t}=\alpha_g+\beta_gP_{g,t}+\delta_gP_{g,t}^2 \tag{10.51}$$

$$E_{g}P_{g,t} = a_{g} + b_{g}P_{g,t} + c_{g}P_{g,t}^{2} \tag{10.52}$$

subject to the following constraints:

$$\sum_{g} P_{g,t} + \sum_{\Re} P_{\Re,t} - \sum_{l \in sl} P_{l,t}^{+} + \sum_{r \in rl} P_{l,t}^{-} + \sum_{y} \alpha_{y,t}^{r} = \sum_{k} P_{k,t} \forall b, t \tag{10.53}$$

$$0 \leq \alpha_{y,t}^{r} \leq \overline{\alpha^{r}} \tag{10.54}$$

$$\overline{\alpha^{r}} \leq P_{k,t} \tag{10.55}$$

$$\sum_{\Re} C_{\Re}^{i} P_{\Re,t}^{\max} \leq UB_{\Re,t} \forall t \tag{10.56}$$

$$0 \leq P_{\Re,t}^{\max} \leq \overline{P_{\Re}} \forall \Re, t \tag{10.57}$$

$$P_{\Re,t}^{\max} = \sum_{c} \alpha_{\Re ct} P_{\Re ct} \forall \Re, t \tag{10.58}$$

$$\sum_{c} \alpha_{\Re ct} = 1 \forall \Re, t \tag{10.59}$$

$$P_{l,t} = B_{l}\left(\phi_{l \in sl,t} - \phi_{l \in rl,t}\right) \forall l, t \tag{10.60}$$

$$-P_{l}^{\max} \leq P_{l,t} \leq P_{l}^{\max} \forall l, t \tag{10.61}$$

$$0 \leq P_{\Re,t} \leq P_{\Re}^{\max} \forall \Re, t \tag{10.62}$$

$$P_{g}^{\min} \leq P_{g,t} \leq P_{g}^{\max} \forall g, t \tag{10.63}$$

$$-\pi \leq \phi_{b,t} \leq \pi \forall b, t. \tag{10.64}$$

The objective function is presented in Eq. (10.50), which seeks to minimize the investment cost of the new renewable generating units, operational cost and emission of fossil fuel-generating units, and the DR costs in the system. Equations (10.51) and (10.52) represent the quadratic cost function for operational cost and emission of the fossil fuel generators. Equation (10.53) is similar to Eq. (10.17), except for the incorporation of optimal DR power output. Equations (10.54) and (10.55) are the DR constraints, which explain that the DR maximum limit must not be exceeded at any given time and that the maximum boundary of the DR must be within the range of the maximum load limit at any time. Constraints (10.56) to (10.64) are similar to Constraints (10.18) to (10.26).

The available time slot for DR on the network is taken as between two and four hours and the average DR was obtained from the percentage of available time with the load at each bus. The DR cost is taken to be \$105/MWh as obtained in [15].

The AIMMS code for solving the GCEP-DR model is given below:

```
Model Main_newwork {
  Set generator {
    Index: g;
  }
  Set nodes {
    Index: n;
  }
  Set demand {
    Index: d;
  }
  Set periods {
    Index: t;
  }
  Set line {
    Index: l;
  }
  Set el {
    SubsetOf: line;
  }
  Set cl {
    SubsetOf: line;
  }
  Set Res {
    Index: R;
  }
  Set Ropt {
    Index: q;
  }
  Set dr {
    Index: rr;
  }
  Parameter mapDrn {
    IndexDomain: (rr,n);
  }
  Parameter Upperdrr {
    IndexDomain: (rr,t);
  }
  Parameter durr {
    IndexDomain: rr;
  }
  Parameter dur2;
  Parameter Costdrr {
    IndexDomain: rr;
  }
  Parameter mapgen {
    IndexDomain: (g,n);
  }
  Parameter mapRes {
    IndexDomain: (R,n);
  }
  Parameter mapdem {
    IndexDomain: (d,n);
  }
  Parameter mapsl {
    IndexDomain: (l,n);
  }
  Parameter maprl {
    IndexDomain: (l,n);
```

```
}
Parameter demands {
  IndexDomain: (d,t);
}
Parameter fmax {
  IndexDomain: l;
}
Parameter bb {
  IndexDomain: l;
}
Parameter pmax {
  IndexDomain: g;
}
Parameter aa {
  IndexDomain: g;
}
Parameter b {
  IndexDomain: g;
}
Parameter c {
  IndexDomain: g;
}
Parameter ea {
  IndexDomain: g;
}
Parameter eb {
  IndexDomain: g;
}
Parameter ec {
  IndexDomain: g;
}
Parameter investl {
  IndexDomain: l;
}
Parameter investR {
  IndexDomain: R;
}
Parameter M;
Parameter dur;
Parameter k;
Parameter kk {
  IndexDomain: t;
  Definition: (1/(1+0.085)^v(t));
}
Parameter v {
  IndexDomain: t;
}
Parameter ResCap {
  IndexDomain: (t,R,q);
}
Parameter URES {
  IndexDomain: (R);
}
Parameter u1;
Parameter u2;
Parameter u3;
Parameter u4;
```

```
Parameter u5;
Variable DRR {
   IndexDomain: (rr,t);
   Range: nonnegative;
}
Variable z {
   IndexDomain: (l,t);
   Range: binary;
}
Variable Rz {
   IndexDomain: (t,R,q);
   Range: binary;
}
Variable RMax {
   IndexDomain: (R,t);
   Range: nonnegative;
}
Variable Rp {
   IndexDomain: (R,t);
   Range: nonnegative;
}
Variable pg {
   IndexDomain: (g,t);
   Range: nonnegative;
}
Variable flow {
   IndexDomain: (l,t);
   Range: nonnegative;
}
Variable ang {
   IndexDomain: (n,t);
   Range: free;
}
Variable Totalcost {
   Range: free;
   Definition: sum(t,(1/(1+k)^v(t))*((u1*sum(R, investR(R)*RMax(R,t)))+(u2*dur*sum(g,
(aa(g)+(b(g)*pg(g,t))+(c(g)*pg(g,t)^2))))+(u3*dur*sum(g,
(ea(g)+(eb(g)*pg(g,t))+(ec(g)*pg(g,t)^2))))+u4*(sum(rr,(dur2*durr(rr)*Costdrr(rr)*DRR(rr,t))))-(u5*0.3*(sum(R,
investR(R)*RMax(R,t))))));
}
MathematicalProgram Leastcost {
   Objective: Totalcost;
   Direction: minimize;
   Constraints: AllConstraints;
   Variables: AllVariables;
   Type: Automatic;
}
Constraint powerbalance {
   IndexDomain: (n,t);
   Definition: sum(g|mapgen(g,n),pg(g,t))+sum(R|mapRes(R,n),Rp(R,t))-
sum(l|mapsl(l,n),flow(l,t))+sum(l|maprl(l,n),flow(l,t))+sum(rr|mapDrn(rr,n),DRR(rr,t))=sum(d|mapdem(d,n),demands(d,t));
}
Constraint Drrlimit1 {
   IndexDomain: (rr,t);
   Definition: 0<=DRR(rr,t);
}
Constraint Drrlimit2 {
   IndexDomain: (rr,t);
```

```
    Definition: DRR(rr,t)<=Upperdrr(rr,t);
}
Constraint linelimit1 {
    IndexDomain: (l,t);
    Definition: -fmax(l)<=flow(l,t);
}
Constraint linelimit2 {
    IndexDomain: (l,t);
    Definition: flow(l,t)<=fmax(l);
}
Constraint genlimit1 {
    IndexDomain: (g,t);
    Definition: 0<=pg(g,t);
}
Constraint genlimit2 {
    IndexDomain: (g,t);
    Definition: pg(g,t)<=pmax(g);
}
Constraint buslimit {
    IndexDomain: (n,t);
    Definition: -3.14<=ang(n,t)<=3.14;
}
Constraint flowlimit1 {
    IndexDomain: (l,t);
    Definition: flow(l,t)=bb(l)*(sum(n|mapsl(l,n),ang(n,t))-sum(n|maprl(l,n),ang(n,t)));
}
Constraint Rlimit {
    IndexDomain: (R,t);
    Definition: RMax(R,t)=sum(q,Rz(t,R,q)*ResCap(t,R,q));
}
Constraint Rlimit2 {
    IndexDomain: (R,t);
    Definition: sum(q,Rz(t,R,q))=1;
}
Constraint Rlimit3 {
    IndexDomain: (R,t);
    Definition: Rp(R,t)<=RMax(R,t);
}
Constraint Rlimit4 {
    IndexDomain: (R,t);
    Definition: 0<=RMax(R,t)<=URES(R);
}
Constraint nn {
    IndexDomain: (R,t)|ord(t)<=9;
    Definition: Rp(R,t+1)>=Rp(R,t);
}
Procedure MainInitialization {
    Comment: "Add initialization statements here that do NOT require any library being initialized already.";
}
Procedure PostMainInitialization {
    Comment: {
        "Add initialization statements here that require that the libraries are already initialized properly,
        or add statements that require the Data Management module to be initialized."
    }
}
Procedure MainExecution {
    Body: {
        solve Leastcost;
    }
}
```

Here, the DR is incorporated into the dynamic GCEP and the optimization results obtained are discussed. Tables (10.17) and (10.18) present the annual optimal power output from the fossil fuel-generating units and the new renewable energy-generating units. It can be seen that as the output power from the new renewable energy-generating units increases across the year, the fossil fuel-generating capacities depreciate. Table 10.18 displays that the new renewable energy-generating unit installed in a particular year cannot be constructed in the next year. Table 10.19 describes the annual OPF on the transmission lines with the presence of DR in the system. Table 10.20 presents the optimal load curtailment owing to 100% DR penetration into the power system.

## 10.7 Summary

This chapter gives detail of a GCEP system. The system provides different planning approaches, ranging from load forecasting, real and reactive power capacity to DR integrated GCEP. The GCEP models were formulated for both DC and AC planning methods, as well as real and reactive power capacity. Different examples are illustrated in this chapter and their solutions are provided with appropriate AIMMS codes.

## 10.8 Nomenclature

The nomenclature used in this chapter is given below:

**Set**

| | |
|---|---|
| g | existing generator |
| k | demand |
| r | demand response |
| $t$ | time |
| $y$ | source of DR |
| $\Re$ | prospective RES |

**Parameters**

| | |
|---|---|
| $C^{i}_{\Re}$ | investment cost of the prospective RES $\Re$ |
| $C^{o}_{g}$ | operational cost of the existing generator g |
| $E_g$ | emissions cost of the existing generator g |
| $\beta$ | duration factor |
| $P_{k,t}$ | system load demand k |
| $P^{min}_{g}$ and $P^{max}_{g}$ | minimum and maximum power capacity of the existing generator g |

**Table 10.17** Optimal power from fossil fuel generators for dynamic GCEP-DR model

| Generators | Bus | Period (year) | | | | | | | | | |
|---|---|---|---|---|---|---|---|---|---|---|---|
| | | 1 | 2 | 3 | 4 | 5 | 6 | 7 | 8 | 9 | 10 |
| G1 | 1 | 150 | 145.36 | 128.43 | 139 | 148.59 | 132.16 | 140.70 | 148.37 | 131.04 | 145.49 |
| G2 | 3 | 6.43 | 30.77 | 68.80 | 115 | 112.47 | 110.30 | 108.14 | 105.97 | 103.80 | 99.47 |
| G3 | 6 | 418.57 | 430.26 | 429.79 | 413 | 420.85 | 425.55 | 430.20 | 433.2 | 436.2 | 440.55 |

**Table 10.18** Optimal power from renewable energy generators for dynamic GCEP-DR model

| Renewable energy sources | Bus | Period (year) | | | | | | | | | |
|---|---|---|---|---|---|---|---|---|---|---|---|
| | | 1 | 2 | 3 | 4 | 5 | 6 | 7 | 8 | 9 | 10 |
| RES 1 | 1 | 125 | 125 | 125 | 125. | 125.00 | 150 | 150 | 150 | 175 | 175 |
| RES 2 | 2 | 40 | 80 | 120 | 158.54 | 193.10 | 200 | 240 | 276.30 | 280 | 320 |
| RES 3 | 4 | – | – | – | – | – | 10 | 20 | 30 | 50 | 90 |

**Table 10.19** Optimal power flow on transmission lines for dynamic GCEP-DR model

| Lines | Period (year) | | | | | | | | | |
|---|---|---|---|---|---|---|---|---|---|---|
| | 1 | 2 | 3 | 4 | 5 | 6 | 7 | 8 | 9 | 10 |
| 1–2 | 21.29 | 17.43 | 11.43 | 13.84 | 16.25 | 17.16 | 18.04 | 18.04 | 18.04 | 17.16 |
| 1–4 | 145.41 | 150 | 150 | 150 | 150 | 150 | 150 | 150 | 150 | 150 |
| 1–5 | 16.80 | 28.32 | 42.98 | 52.04 | 61.10 | 64.50 | 67.80 | 67.80 | 67.80 | 64.50 |
| 2–4 | 38.51 | 41.10 | 45.00 | 55.50 | 54 | 52.50 | 51 | 49.50 | 48.00 | 45 |
| 2–6 | 79.59 | 97.50 | 120 | 130.50 | 129 | 127.50 | 126 | 124.50 | 123.00 | 120 |
| 3–5 | 146.66 | 150 | 150 | 150 | 150 | 150 | 150 | 150 | 150 | 150 |
| 3–6 | 150 | 150 | 142.20 | 121.20 | 124.20 | 127.20 | 130.20 | 133.20 | 136.20 | 142.20 |
| 4–6 | 121.91 | 130.26 | 137.59 | 142.12 | 146.65 | 148.35 | 150 | 150 | 150 | 148.35 |

**Table 10.20** Optimal load curtailment for 100% participating factor for dynamic GCEP-DR model

| Demand Response | Available time slot, hr | Period (year) | | | | | | | | | |
|---|---|---|---|---|---|---|---|---|---|---|---|
| | | 1 | 2 | 3 | 4 | 5 | 6 | 7 | 8 | 9 | 10 |
| 1 | 2 | 3.33 | 3.67 | 4.00 | 4.33 | 4.67 | 5.00 | 5.33 | 5.67 | 6.00 | 6.67 |
| 2 | 3 | – | 2.62 | 15.98 | – | – | 22.00 | 10.80 | – | 21.80 | 29.50 |
| 3 | 2 | 1.67 | 1.83 | 2.00 | 2.17 | 2.33 | 2.50 | 2.67 | 2.83 | 3.00 | 3.33 |
| 4 | 4 | – | – | – | – | – | – | 0.16 | 6.16 | 2.16 | – |
| 5 | 3 | 15 | 16.5 | 18 | 19.5 | 21 | 22.5 | 24 | 25.5 | 27 | 30 |

$\overline{P_{\Re}}$ upper bound of prospective RES $\Re$
$\delta_y$ available time slot at DR source $y$
$C^{r}_{y,t}$ cost of power curtailed by DR $r$ at source $y$ in time $t$
$\overline{\alpha^{r}}$ maximum capacity value of DR $r$
$UB_{\Re,t}$ utility budget for prospective RES $\Re$ at time $t$
$P_{\Re ct}$ generation capacity of RES $\Re$ at investment choice c in time $t$
$B_l$ susceptance of transmission line

## Variables

| | |
|---|---|
| $P_{\Re,t}^{\max}$ | capacity of prospective renewable energy source $\Re$ at time $t$ |
| $P_{g,t}$ | power produced by the existing generator g at time $t$ |
| $P_{\Re,t}$ | power produced by the RES $\Re$ at time $t$ |
| $P_{l,t}$ | power flowing on transmission line l at time $t$ |
| $P_{l,t}^{+}$ and $P_{l,t}^{-}$ | power flowing into and out of the transmission line l at time |
| $\alpha_{y,t}^{r}$ | active power curtailed by DR $r$ at source $y$ in time $t$ |
| $\alpha_{\Re ct}$ | binary variable for choosing prospective RES $\Re$ at option c at time $t$ |
| $\phi_{b,t}$ | voltage angle at bus $b$ at time $t$ |

# References


1. Nwulu NI, Xia X (2015) Multi-objective dynamic economic emission dispatch of electric power generation integrated with game theory based demand response programs. Energy Convers Manag 89:963–974
2. Zhang N, Hu Z, Springer C, Li Y, Shen B (2016) A bi-level integrated generation-transmission planning model incorporating the impacts of demand response by operation simulation. Energy Convers Manag 123:84–94
3. Rouhani A, Hosseini SH, Raoofat M (2014) Composite generation and transmission expansion planning considering distributed generation. Int J Electr Power Energy Syst 62:792–805
4. Hemmati R, Hooshmand RA, Khodabakhshian A (2013) Reliability constrained generation expansion planning with consideration of wind farms uncertainties in deregulated electricity market. Energy Convers Manag 76:517–526
5. Gbadamosi SL, Nwulu NI (2020) A multi-period composite generation and transmission expansion planning model incorporating renewable energy sources and demand response. Sustain Energy Technol Assess 39 (October 2019), 100726
6. Gbadamosi SL, Nwulu NI, Sun Y (2018) Multi-objective optimisation for composite generation and transmission expansion planning considering offshore wind power and feed-in tariffs. IET Renew Power Gener 12(14):1687–1697
7. Gbadamosi SL, Nwulu NI (2020) Optimal power dispatch and reliability analysis of hybrid CHP-PV-wind systems in farming applications. Sustainability 12 (2020)
8. Gbadamosi SL, Nwulu NI, Damisa U (2019) Impact of power losses on optimal expansion planning using quadratic programming. In: 6th IEEE international conference on engineering technologies and applied sciences (ICETAS), pp 1–4
9. Gbadamosi SL, Nwulu NI (2019) A dynamic optimal expansion planning model integrating renewable energy sources. In: 11th international conference on applied energy, pp 150–155
10. Gbadamosi SL, Nwulu NI (2020) Optimal planning of renewable energy systems for power loss reduction in transmission expansion planning. J Eng Des Technol 18(5):1209–1222
11. Zhang J, Wei Y, Li D, Tan Z, Zhou J (2018) Short term electricity load forecasting using a hybrid model. Energy 158:774–781
12. Gbadamosi SL, Nwulu NI (2020) Reliability assessment of composite generation and transmission expansion planning incorporating renewable energy sources. J Renew Sustain Energy 026301(12)
13. Gbadamosi SL, Nwulu NI, Sun Y (2018) Harmonic and power loss minimization in power systems incorporating renewable energy sources and locational marginal pricing. J Renew Sustain Energy 10:1–24

14. Gbadamosi SL, Nwulu NI, Sun Y (2018) Optimal scheduling in offshore wind farms considering harmonic losses and locational marginal pricing. In: 2018 IEEE PES-IAS 5th Power Africa Conference, pp 658–663
15. Hajebrahimi A, Abdollahi A (2017) Probabilistic multiobjective transmission expansion planning incorporating demand response resources and large-scale distant wind farms. IEEE Syst J 11(2):1170–1181

# Chapter 11
# Conclusion and Future Work

## 11.1 Conclusion

The book *Optimal Operation and Control of Power Systems Using an Algebraic Modelling Language* is motivated by a desire to present an easy-to-read comprehensive text for power system operators and planners to control today's complex grids optimally. This is imperative in the fast-changing context of the power system, especially with customers who are active participants and can curb/control their demand.

AMLs provide an opportunity for modelers to ignore complications in the modeling process, minimize human effort and time and represent optimization models using high-level syntax and representation close to how modelers use mathematical symbolization. These languages also allow for splitting of the mathematical model, solvers, and the data, thus facilitating extensive experimentation.

In this book, AIMMS, which is one of the top five AMLs, was used to model and solve power system operational and control problems.

Chapters 1–3 introduced the various key subjects in the book. Chapter 1 provided an overview of the power system and RES and highlighted topical issues such as the smart grid and the deregulation of power systems. Chapter 2 delved into DSM programs spanning energy efficiency and demand response programs. Both types of DR programs, price-based and IB-DR programs, were considered. Chapter 3 introduced mathematical optimization programs specifically zeroing in on AML and then AIMMS.

Chapters 4–10 covered operation and control issues in power systems spanning various periods. Thus, the dynamic economic emissions dispatch problem, generator maintenance scheduling, optimal power flow, hybrid grid-connected renewable energy powered microgrid, transmission expansion planning and generator expansion planning were modeled and solved using AIMMS. An IB-DR program, the game theory demand response program, integrated with dynamic economic



N. Nwulu and S. L. Gbadamosi, *Optimal Operation and Control of Power Systems Using an Algebraic Modelling Language*, Green Energy and Technology,
https://doi.org/10.1007/978-3-030-00395-1_11

emissions dispatch, combined heat and power dynamic economic emissions dispatch, and hybrid grid-connected renewable energy microgrid, among others, were also modeled and solved using AIMMS. The source AIMMS codes for all these examples were presented and detailed.

There are a number of avenues available for extension of the work presented in this book. Firstly, further advances in today's grid, inclusive of vehicle-to-grid and grid-to-vehicle among other electric vehicle setups, can be modeled using AIMMS. The advent of prosumers also requires further investigation using AML platforms. There is also a need to investigate increased integration of AMLs into more advanced power system deregulated setups.

Furthermore, digital technologies, tools, and techniques, such as IoT, blockchain, artificial intelligence, machine learning [1–3], and big data can be integrated and hybridized with AMLs for improved power system operations. For example, machine-learning techniques can be used for classification, clustering, prediction, forecasting, and regression of key power system parameters, which can further be fed as data into AML models.

Other possible extensions from a modeling perspective include the consideration of AMLs for model predictive control [4], switched model predictive control, robust optimization [5], stochastic mathematical modeling [6], logical discrete optimization, and bi-level optimization, among others.

## References

1. Nwulu NI, Fahrioglu M (2011) A neural network model for optimal demand management contract design. In: Proceedings of the 10th international conference on environmental and electrical engineering (EEEIC2011), Rome, Italy, 8–11 May 2011
2. Nwulu NI, Agboola PO (2012) Modelling and predicting electricity consumption using artificial neural networks. In: Proceedings of the 11th international conference on environmental and electrical engineering (EEEIC2012), Venice, Italy, 18–25 May 2012
3. Nwulu NI, Fahrioglu M (2011) Power system demand management contract design: a comparison between game theory and artificial neural networks. Int Rev Model Simul 4 (1):104–112
4. Nwulu NI, Xia X (2015) Implementing a model predictive control strategy on the dynamic economic emission dispatch problem with game theory based demand response programs. Energy 91:404–419
5. Damisa U, Nwulu NI, Sun Y (2018) A robust energy and reserve dispatch model for prosumer microgrids incorporating demand response aggregators. J Renew Sustain Energy 10(5)
6. Ioannou A, Fuzuli G, Brennan F, Yudha SW Angus A (2019) Multi-stage stochastic optimization framework for power generation system planning integrating hybrid uncertainty modelling. Energy Econ 80:760–776

The manufacturer's authorised representative in the EU is Springer Nature Customer Service Centre GmbH, Europaplatz 3, 69115 Heidelberg, Germany. If you have any concerns regarding our products, please contact ProductSafety@springernature.com

Printed and bound by CPI Group (UK) Ltd, Croydon, CR0 4YY
30/09/2026
02225440-0006